耕地细碎化与农村集体行动：
以中国农田灌溉系统为例

FARMLAND FRAGMENTATION AND COLLECTIVE ACTION:
A Study on the Irrigation System in China

臧良震 ◎著
Liangzhen Zang

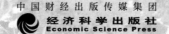

ACKNOWLEDGEMENT

This book is jointly supported by the Agricultural Science and Technology Innovation Program (ASTIP-IAED-2021-08), the National Natural Science Foundation of China (71573151 & 71721002), the National Social Science Fund of China (15ZDB164), the National Key Research Program (2016YFC0401408), the Fundamental Research Funds for the Central Public Interest Scientific Institute of China (161005202118). The data used in this book is supported by China Institute for Rural Studies, Tsinghua University (CIRS).

PREFACE

The tragedy of the commons is a classic problem faced by population, resources and environmental management. Since Garrett Hardin published the *Tragedy of the Commons* on *Science* in 1968, a new discipline direction of governing the commons has emerged in academic discussion. After more than half a century, the research on governing the commons has attracted many scholars worldwide and formed a complete knowledge system. Effective collective action is the key to solving the problem of the tragedy of the commons, and collective action theory has thus become the essence of the knowledge system in governing the commons.

Since the new century, the phenomenon of paradox has appeared in many rural areas of China: with the development of rural economy, the growth of farmers' income, the increase of agricultural production yield and the continuous improvement of infrastructure, the governance of the commons in rural areas such as ecological environment, farmland water conservancy and humanistic environment are generally declining. The essence of China's rural governance crisis is that under the background of the transformation of rural social structure and the rapid change of rural system, the governance of rural commons has changed from collectivization to individualization, resulting in the overall decline of collective action in rural China. A large number of research literatures, focusing on how to effectively facilitate the collective action, have formed a systematic theoretical understanding of the important factors and mechanisms affecting collective action in rural China.

As the basic resource of agricultural production, farmland plays an important role in governing the commons. Smallholder agriculture is the foundation of

FARMLAND FRAGMENTATION AND COLLECTIVE ACTION:
A STUDY ON THE IRRIGATION SYSTEM IN CHINA

food security and an important part of the socio-ecological system in many countries. According to the FAO's World Census of Agriculture, the proportion of farm area less than 1ha, 2ha and 5ha are 73%, 85% and 95% respectively in 81 countries (2/3 of the world population and about 38 percent of the arable lands worldwide). In terms of distribution, smallholders are common in Europe, OECD countries, and developing countries such as Brazil, India, and China. Therefore, discussing the role of farmland fragmentation in governing the commons is of great significance to countries with a small area of arable land per capita in the world, especially for developing countries, to promote agriculture and rural development. It also constitutes an important topic in the study of public administration and collective action theory.

In the above background, this book authored by Dr. Liangzhen Zang focuses on the relationship between farmland fragmentation and collective action, which is a new contribution in the academic direction of governing the commons. Dr. Zang examined an interesting question: What is the impact of farmland fragmentation, a variable with Chinese characteristics, on collective action? This research question has important theoretical value in governing the commons, for a large country with abundant smallholders is China's national and agricultural situation and such an examination shed new lights to the logic of collective action in the commons in rural China. Taking the irrigation commons as an example, Dr. Zang reveals some new mechanisms of farmers' collective action, and thereby enriches our knowledge around collective action and helps us understand how to deal with the dilemma in governing the commons in rural China. I congratulate him on this innovative work as well as the publication of this book.

Professor and Associate Dean, School of Public Policy and Management at Tsinghua University
Vice President, China Institute for Rural Studies at Tsinghua University

ABSTRACT

A big country with a large population of smallholders is the fundamental realities of China and has a decisive influence on the success or failure of rural collective action. Based on the farmland irrigation system, this book starts from the perspective of small-scale peasants' farmland fragmentation, and uses Institutional Analysis and Development (IAD) framework and Social-ecological System (SES) framework to discuss the logic of small-scale peasants' participation in the rural irrigation collective action. It analyzes the impact of farmland fragmentation on rural irrigation collective action, reveals the mechanism and makes an in-depth analysis of the causes of the impact. The results show that the degree of farmland fragmentation in China is relatively high, which seriously restricts the formation of rural irrigation collective action. Although it is difficult to change the negative effect of farmland fragmentation on rural irrigation collective action, effective institutional rules can weaken the negative effect of farmland fragmentation on collective action to a certain extent. The effect of farmland fragmentation on the rural irrigation collective action in China is not shown directly, but plays a role through different mediating factors. Specifically, farmland fragmentation reduces peasants' dependency on agricultural production, restrains the formation of irrigation rules, and increases the economic pressure faced by peasants when they engage in rural irrigation collective action, thus resulting in the overall decline of China's rural irrigation collective action. On the one hand, this book enriches the existing theories related to collective action, especially making an important supplement to IAD framework and SES framework, which has important theoretical value. On the other hand, this book reveals the internal mechanism of farmland fragmentation affecting rural irrigation collective action, which exerts important guiding significance on the overall improvement of rural collective action capacity in the future.

CONTENTS

LIST OF TABLES

LIST OF FIGURES

CHAPTER 1 INTRODUCTION 1
    1.1   Background and Questions 1
        1.1.1   Background 1
        1.1.2   Questions 4
    1.2   Research Purposes and Significance 6
        1.2.1   Research Purposes 6
        1.2.2   Significance 7
    1.3   Structure of This Book 9
    1.4   Methodology and Technology Roadmap 11
        1.4.1   Methodology 11
        1.4.2   Technical Roadmap 12

CHAPTER 2 LITERATURE REVIEW AND THEORETICAL BASIS 13
    2.1   Literature Review 13
        2.1.1   Farmland Fragmentation 13
        2.1.2   Rural Irrigation Collective Action 51
    2.2   Theoretical Basis 59
        2.2.1   Institutional Analysis and Development Framework 59

2.2.2　Social-ecological System Framework　　　　62

CHAPTER 3　LOGICAL RELATIONSHIP BETWEEN FARMLAND RESOURCES AND RURAL IRRIGATION COLLECTIVE ACTION　65

3.1　Collective Action of Small-scale Peasant Households　　66

    3.1.1　Historical Evolution of Participation of Small-scale Peasant Households in Rural Collective Action　　66

    3.1.2　Logical Relationship between the Characteristics of Small-scale Peasant Households and Rural Collective Action　69

    3.1.3　Logic of Small-scale Peasant Households' Participation in Rural Collective Action based on IAD Framework　73

3.2　Farmland Resources and Rural Irrigation Collective Action　74

    3.2.1　Farmland Size and Rural Irrigation Collective Action　74

    3.2.2　Farmland Location and Rural Irrigation Collective Action　76

    3.2.3　Farmland Property Rights and Rural Irrigation Collective Action　77

    3.2.4　Logical Relationship between Farmland Resources and Rural Collective Action based on IAD Framework　77

3.3　Logical Relationship between Farmland Fragmentation and Rural Irrigation Collective Action　78

3.4　Conclusion　　80

CHAPTER 4　IMPACT OF FARMLAND RESOURCE ENDOWMENT ON RURAL IRRIGATION COLLECTIVE ACTION　82

4.1　Data　　83

4.2　Model　　84

4.3　Results　　86

    4.3.1　Descriptive Analysis　　86

    4.3.2　Results of Econometric Analysis　　87

4.4　Conclusion　　89

CHAPTER 5 TOTAL EFFECT OF FARMLAND FRAGMENTATION ON RURAL IRRIGATION COLLECTIVE ACTION   91

    5.1 Framework   92

        5.1.1 IAD Framework   92

        5.1.2 Selection of Influencing Factors and Research Hypothesis   93

    5.2 Sample Selection and Method   99

        5.2.1 Sample Selection   99

        5.2.2 Method   99

    5.3 Results   101

        5.3.1 Descriptive Analysis   101

        5.3.2 Results of Econometric Analysis   103

    5.4 Discussion   112

    5.5 Conclusion   114

CHAPTER 6 MEDIATING EFFECT OF FARMLAND FRAGMENTATION ON RURAL IRRIGATION COLLECTIVE ACTION   116

    6.1 Framework and Potential Mechanisms   117

        6.1.1 SES Framework   117

        6.1.2 Analysis of Potential Mechanism   117

    6.2 Hypotheses   121

    6.3 Method, Variables and Model   122

        6.3.1 Method   122

        6.3.2 Variables   123

        6.3.3 Model   126

        6.3.4 Samples   127

    6.4 Total Effect of Farmland Fragmentation on Collective Action   128

    6.5 Discussion   131

        6.5.1 Effects of Dependency on Farming   131

        6.5.2 Effects of Irrigation Rule-making   132

	6.5.3	Effects of Economic Pressure	133
	6.5.4	Effects of Land Circulation	133
6.6	Conclusion		134

CHAPTER 7　CONCLUSIONS, POLICY RECOMMENDATIONS AND PROSPECTS　136

7.1	Conclusions	136
7.2	Policy Recommendations	138
7.3	Prospects	140
	7.3.1　Innovations	140
	7.3.2　Deficiency and Prospect	142

REFERENCE　144

LIST OF TABLES

Table 2–1　Classification of goods / 61

Table 2–2　Social-ecological system(Second layers) / 63

Table 4–1　Descriptive results of village-level variables / 87

Table 4–2　Farmland resource endowment influences the measurement results of rural irrigation collective action / 87

Table 5–1　Variables definition and expected impact on irrigation collective action / 100

Table 5–2　Descriptive statistics / 103

Table 5–3　The determinants of participation in construction and maintenance of collective irrigation / 104

Table 5–4　The determinants of attending meeting of collective irrigation / 105

Table 5–5　Results of peasants' participation in the construction and maintenance of collective irrigation facilities (with interaction) / 109

Table 5–6　Results of peasants' participation in collective irrigation facilities related meetings (with interaction) / 110

Table 6–1　Decomposition of Social-ecological System framework / 121

Table 6–2　Reliability statistics of the latent variables / 128

Table 6–3　Result of direct and indirect effect of land fragmentation on irrigation collective action / 129

LIST OF FIGURES

Figure 1-1　Technical roadmap　12

Figure 2-1　Institutional Analysis and Development framework　60

Figure 2-2　Rules as exogenous variables directly affecting the elements of an action situation　62

Figure 2-3　Social-ecological system (First layers)　63

Figure 3-1　Logical relationship between the characteristics of small-scale peasant households and rural collective action based on IAD framework　73

Figure 3-2　Logical relationship between farmland resources and rural collective action based on IAD framework　78

Figure 5-1　Logical relationship between farmland fragmentation and rural irrigation collective action based on IAD framework　93

Figure 6-1　Mechanisms mediating the influence of farmland fragmentation on irrigation collective action　131

Figure 7-1　The logic diagram of farmland fragmentation affecting rural irrigation collective action　138

CHAPTER 1

INTRODUCTION

1.1 Background and Questions

1.1.1 Background

Since the reform and opening up in 1978, China's agriculture and rural areas have undergone earth-shaking changes. The governance of the commons, however, is faced with increasingly severe problems, such as the deterioration of rural human settlements, the degradation of eco-environment, and the insufficient supply of public services, so that it becomes difficult to maintain collective actions in rural areas (Wang, 2017). That is because it is inextricably linked to China's fundamental realities as a big country with a large population of smallholders. Smallholders refer to the microcosmic agricultural entity that integrates production and consumption based on households. According to the *Third National Agricultural Census Data Bulletin*, as of 2016, among the 207 million agricultural households across China, smallholders accounted for a high proportion of 98.1%. Smallholders serve as the most fundamental unit of agricultural production in China, and also the most fundamental force that plays a major role in the governance of the commons in rural areas of China.

Xi Jinping's report at 19*th CPC National Congress* clearly proposed to implement the strategy for rural revitalization, and specified the general requirements of Prosperous Industry, Livable Ecology, Civilized Rural Customs, Effective Governance, and Living in Abundance. In such a big country with a large population of smallholders, the key to achieving the goal of Effective Governance is to promote peasant households to engage in the governance of rural public affairs, especially to solve the current problem of insufficient supply of rural public goods. Therefore, how to effectively encourage peasant households to engage in the supply of rural public goods has become the top priority in solving the tough problem of insufficient supply of rural public goods and imbalance between supply and demand. Common-pool Resources (CPR) including collective irrigation facilities are an important part for the governance of the commons. It is significant for realizing effective rural governance to clarify the collective actions logic of peasant households engaging in the governance of the type of commons (Cai, 2017; Qiao, 2016).

The farmland irrigation facilities are the typical common-pool resources, because the construction and maintenance of irrigation facilities directly reflect the capacity of taking collective action in rural areas. In terms of the development of farmland irrigation facilities in China, the basic unit of rural collective actions for irrigation has been narrowed from the initial people's communes to villages, and in recent years, further narrowed to a few peasant households or a single peasant household. That is, the governance of the commons has shifted from collective cooperation to individual cooperation, and the capacity of taking rural irrigation collective action has witnessed an overall decline. Theoretically speaking, peasant households' participation in irrigation collective action is a process in which many geographically-based individuals make independent decisions about whether to participate under the guidance of the government. As a matter of fact, however, the rivalry and non-excludability of rural irrigation facilities led to inconsistencies between individual and collective decisions, causing troubles to collective actions.

At present, many scholars have carried out researches on the topic of factors influencing rural irrigation collective actions, such as village size (Jacob, 2016; Poteete, 2008; Cai, 2014), income heterogeneity (Cai, 2016), social network (Aymen, 2017), reward and punishment mechanism (Totin, 2014), social capital (Jaime, 2013; Hoogesteger, 2013), labor force (Wang, 2017; Aatika, 2016; Nagrah, 2016), operation and management methods (Jyoti, 2016; Aymen, 2017; Frija, 2017), land scale (Chun, 2014), water user associations (Jaime, 2015; Hoogesteger, 2015), rules-in-use (Totin, 2014; Jennewein, 2016; Patt, 2017), etc. On the whole, all influencing factors can be generalized into three attributes, that is, biophysical conditions, attributes of community, and rules-in-use (Ostrom, 1994; Araral, 2008). Among them, the biophysical conditions, as the realistic basis for the implementation of rural irrigation collective action, can also affect the impact of attributes of community and rules-in-use on collective action to a certain extent. Therefore, it is particularly essential to study the impact of biophysical conditions on rural irrigation collective actions.

Farmland, as the most basic means of agricultural production, plays an important role in rural irrigation collective action. In China, due to the long-term existence of the urban-rural dual economic structure, many smallholders have developed a strong dependence on land resources. In addition, China's farmland resource endowment per capita suffers from a long-term shortage. According to the data published by the World Bank, in 2014, the farmland area per capita in the world recorded 0.196 *ha*, and the farmland area per capita in countries like Australia and the United States reached 2.002 *ha* and 0.485 *ha* respectively. In comparison, the farmland area per capita in China was only 0.077 *ha*, far behind the global average and the levels of major developed countries in the world. Even compared with the condition in India, whose population and territory area are relatively close to China, there is still a gap from their level of 0.121 *ha*. Moreover, after the implementation of the Household Contract Responsibility System in China, the farmland has become severely fragmented (Lyu, 2011). The average farmland area of rural households is less than 8 *mu*, which is divided into 4 to 5 plots. It can be seen that the study of the impact of farmland fragmentation in China on rural irrigation collective action is of important practical significance for solving the problem of effective supply of rural collective irrigation facilities.

1.1.2 Questions

Based on the above background, the following questions are proposed in this book.

First, how does smallholders' endowment of farmland resources affect the rural irrigation collective action? According to the *National Population Development Plan* (2014 – 2020), even if China realizes an urbanization rate of 70% by 2030, more than 400 million people will still live in rural areas, and the problem of insufficient resource endowments will be hardly changed in the long term. Based on this practical problem, it is significant to analyze the impact of farmland resource endowment on rural irrigation collective action. Will China's insufficient endowment of farmland resources affect the capacity of rural irrigation collective action? If it does, will it have an inhibitory effect? Mastering the impact of farmland resource endowments on rural irrigation collective action is conducive to deeper analysis and exploration of the impact mechanism of farmland fragmentation on rural irrigation collective action.

Second, what is the impact of farmland fragmentation on rural irrigation collective action, and what are the effects of different rules-in-use? Will farmland fragmentation exert a certain impact on the rural irrigation collective action? If it does, will the impact be positive or negative? Under the Institutional Analysis and Development (IAD) framework, it is difficult to change the biophysical conditions such as the abundance of water resources and village distance, as well as the attributes of community such as the economic development and the size of the village. In contrast, the rules-in-use can be changed more flexibly and acts on the rural irrigation collective action. Therefore, can the current rules-in-use moderate the impact of farmland fragmentation on the rural irrigation collective action? If it can, what are the effects of different rules-in-use? These questions will be discussed one by one in this book.

Third, what are the intermediary factors through which farmland fragmentation affects the rural irrigation collective action? If farmland fragmentation does have a certain impact on the rural irrigation collective action, will the total effect be manifested through a direct effect? Or are there various intermediary factors, and farmland fragmentation affects the rural irrigation collective action through the action of those various intermediary factors? What's more, how can we locate these intermediary factors? If these factors exist, which factors have a positive effect, and which factors have a negative effect? What is the overall manifestation? By answering the above questions, we will be able to fully understand the mechanism of how farmland fragmentation affects rural irrigation collective action.

1.2 Research Purposes and Significance

1.2.1 Research Purposes

In this book, based on the current unique phenomenon of severe farmland fragmentation in China, the author analyzed the impact of farmland resource endowment on rural irrigation collective action from both macro and micro perspectives, and, on this basis, judged the relationship between farmland fragmentation and the current rules-in-use, and analyzed the moderating effect of the rules-in-use between farmland fragmentation and rural irrigation collective action. Besides, by introducing intermediary variables, this book explored the moderating mechanism of the impact of farmland fragmentation on rural irrigation collective action, and found out the interaction relationship of different intermediary factors between the two. By studying the mechanism of how farmland fragmentation affects rural irrigation collective action, this book has provided basis for scientific decision-making for enhancing the capacity of rural irrigation collective action, and thus to promote the governance of the commons and achieve the goal of the Effective Governance for rural revitalization in China.

Specifically, this book mainly includes the following objectives. The first purpose is to clarify the mechanism by which farmland resource endowment affects rural irrigation collective action, to explore the historical evolution of China's farmland fragmentation and rural irrigation collective action from the starting point of farmland resources, figure out the facts of the two, and find out the interaction mechanism between the two as a whole. The second purpose is to grasp the direct impact of farmland fragmentation on rural irrigation collective action and the moderating effect of different rules-in-use between the two. On the basis of qualitative analysis, by constructing an econometric model, this book will analyze the overall effect of farmland fragmentation on rural irrigation collective action, and judge whether farmland fragmentation promotes or inhibits the success of rural irrigation collective action in China. On this basis, by introducing a regulatory mechanism, we should grasp the impact of different rules-in-use on the relationship between the two. The third purpose is to refine the potential intermediary factors that affect the relationship between farmland fragmentation and rural irrigation collective action to explore the interaction mechanism between the two with the aid of an econometric analysis model, and to clarify the effects of different mechanisms.

1.2.2 Significance

China is now witnessing a rapid development in agriculture and rural areas, with significantly increased income for peasants. At the same time, however, due to the imbalance of urban and rural development and the poor natural resource endowment, there is great pressure on the supply of rural public goods. The capacity of rural collective action in recent years, especially showed a continuous downward trend, and the supply of rural public goods, exemplified by irrigation facilities, even directly reflects the status quo of governance of rural public affairs in China. From the perspective of China's unique farmland fragmentation, exploring the impact mechanism of farmland fragmentation on rural irrigation collective action is of important theoretical and practical significance.

From a theoretical point of view, first, the analysis in this research is based on the two important frameworks for the study of collective action,

FARMLAND FRAGMENTATION AND COLLECTIVE ACTION: A STUDY ON THE IRRIGATION SYSTEM IN CHINA

Institutional Analysis and Development (IAD) and Social-ecological System (SES), and starts with an external variable with Chinese characteristics, farmland fragmentation, to greatly enrich these two research frameworks as an important supplement to current theories. Second, this research combines the theoretical knowledge of sociology, economics, public management and several other disciplines, and collects materials and accumulates data through field investigations, to develop a theoretical system of the relationship between resources and the governance of the commons, which will be an important theoretical supplement to the research on the governance of the commons.

From a practical point of view, first, this book can provide a corresponding scientific basis for the formulation of China's policies for the governance of the commons in rural areas in the future. This book not only revealed the direct impact of farmland fragmentation on rural irrigation collective action, but also analyzed the moderating effect of different rules-in-use. Meanwhile, this book analyzed and accounted for the indirect effects of farmland fragmentation on rural irrigation collective action. The analysis of the reasons behind and the discussion of the role of different policy rules provide an important scientific basis for the effective governance of China's rural communities in the future, and an important decision-making reference for policy formulation in the future. Second, this book can provide new ideas for the realization of the goal of Effective Governance in rural areas. The purpose of promoting collective action in rural areas is to achieve effective rural governance. As an important input for agricultural production, farmland resources play an important role in the governance of the commons, especially in the process of supplying public goods including rural production facilities. Different attributes of farmland resources jointly determine the success of collective actions. Therefore, it will provide important scientific references for effective governance of rural communities in the future by judging the impact of farmland fragmentation on the governance of the commons, especially by identifying the impact of different intermediary factors in the process.

1.3 Structure of This Book

This book takes China's rural irrigation collective action as the research object, and farmland fragmentation as the perspective, to explore the relationship between farmland fragmentation and the rural irrigation collective action. In general, this book mainly contains the following four parts.

The first part contains two chapters. Chapter 1 is the Introduction. In this chapter, on the basis of explaining the basic background of China's current farmland fragmentation and rural irrigation collective action, the author clarifies the research purpose, expounds the research significance, and specifically describes the research methodology and roadmap. Chapter 2 is the Literature Review and Theoretical Basis. In this chapter, the author mainly sorts out the concept, characteristics, causes, evaluation methods, and impact effects of farmland fragmentation, and summarizes the status quo and influencing factors of rural irrigation collective action, so as to locate the deficiencies of the extant research. The author also sorts and analyzes the main points of view as supports for the follow-up analysis of this research. In addition, this chapter also includes interpretations on IAD and SES frameworks.

The second part is Chapter 3, which is mainly a qualitative analysis between farmland resources and rural irrigation collective action. Specifically, this part mainly probes into the impact of the fundamental realities of a big country with a large population of smallholders on rural irrigation collective action from the perspective of smallholders, and sorts out the historical evolution and the relationship between smallholders and rural irrigation collective action. On this basis, focusing on the object of farmland resources, this part discusses the impact of different attributes of farmland, such as farmland size, location, and ownership, on rural irrigation collective action. The important research object of farmland fragmentation is thereby introduced, and the possible impact of farmland fragmentation on rural irrigation collective action is discussed from a qualitative perspective.

The third part is an analysis of the impact of the farmland resource endowment and farmland fragmentation on the rural irrigation collective action mainly from a quantitative perspective, which contains three chapters. Chapter 4 is to discuss the impact of farmland resource endowment on rural irrigation collective action from the perspective of the village level by adopting an econometric model. Chapter 5 provides an analysis based on IAD framework from the perspective of smallholders, mainly using village-level and smallholder-level data, and an overall analysis of whether farmland fragmentation would have an impact on rural irrigation collective action through the construction of an ordered Probit model. At the same time, by constructing a moderating effect model this chapter also analyzes the moderating relationship of different rules-in-use between the two. In addition, Chapter 6 is a further study based on SES framework of the intermediary mechanism of the impact of farmland fragmentation on rural irrigation collective action, by constructing a structural equation model.

The fourth part is Chapter 7, which is the conclusions, policy recommendations and prospects. This part further summarizes the main conclusions of this book as a whole, and proposes relevant policy recommendations and suggestions to promote rural collective action in China based on the research findings.

1.4　Methodology and Technology Roadmap

1.4.1　Methodology

This book mainly adopts the following methods.

First, the literature summarization. By consulting the extant researches, this book focused on sorting out the characteristics, causes, and evaluation methods of farmland fragmentation, summarized and classified the impact of farmland fragmentation and the reasons behind it, sorted out the theories, frameworks, and influencing factors of collective action, and made literature analysis mainly in the field of rural irrigation collective action. Through literature summarization, on the one hand, the author gained a new entry point for this book to analyze the mechanism underlying farmland fragmentation and rural irrigation collective action; on the other hand, it provides a theoretical reference for the exploration of specific mechanisms.

Second, the qualitative analysis. At present, literatures on the impact of farmland fragmentation and the external influencing factors of rural irrigation collective action present different conclusions, and because the external influencing factors of rural irrigation collective action are characterized by diversity and complexity, how to find the intermediary mechanism between the two and how to analyze the comprehensive impact of different mechanisms on rural irrigation collective action will be a key problem that should be solved. To this end, this book starts from the impact of farmland fragmentation based on literature review, and qualitatively analyzes the possible influencing mechanisms between the two under IAD framework, to provide support for the subsequent quantitative analysis.

Third, the quantitative analysis. The quantitative analysis of this book mainly contains three parts. Firstly, this book analyzed the impact of farmland resource endowment on rural irrigation collective action based on village-level data from a macro perspective by constructing an econometric model. Secondly, to explore the overall effect of farmland fragmentation on the rural irrigation collective action, this book also adopted an ordered Probit method, and added intermediary variables of moderating effect to analyze the moderating effect of different rules-in-use between farmland fragmentation and the rural irrigation collective action. Thirdly, to further analyze the moderating effect of farmland fragmentation on rural irrigation collective action, this book also provided analysis through constructing a structural equation model.

1.4.2 Technical Roadmap

The specific technical roadmap of this book is shown in Figure 1–1.

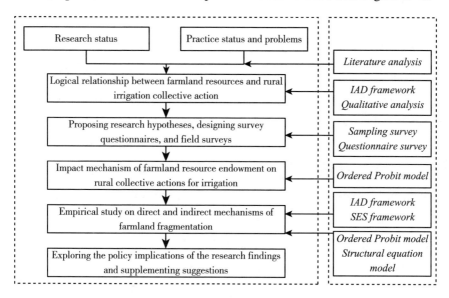

Figure 1–1 Technical roadmap

CHAPTER 2

LITERATURE REVIEW AND THEORETICAL BASIS

2.1 Literature Review

2.1.1 Farmland Fragmentation

The farmland fragmentation is one of the main factors restricting the scale of agricultural operations, and also a difficult problem facing the sustainable development of agriculture. The phenomenon of farmland fragmentation can be found widespread in all countries around the world, especially in Central and Eastern Europe, India, China and other regions and countries (Sklenicka, 2016). Multiple factors including institutional rules, population and culture lead to the phenomenon of farmland fragmentation. Therefore, the impact of farmland fragmentation is also manifested in different ways (Lu, 2017). At present, many scholars have carried out many researches on the problem of farmland fragmentation and have achieved fruitful results (Sklenicka, 2017; Yucer, 2016; Janus,

2017; Bai, 2014; Lu, 2017). However, due to the diversity and complexity of the origins, evaluation methods, and effects of farmland fragmentation, extant researches present various viewpoints. In this part, the author attempts to systematically sort and summarize the current Chinese and foreign literatures from such aspects as connotation, characteristics, origins, evaluation methods, and impact effects of farmland fragmentation, for the purpose of summarizing the research perspectives and viewpoints on the issue of farmland fragmentation. Besides, this part focuses on revealing the mechanism of various effects of farmland fragmentation from the theoretical perspective, in order to provide more theoretical support for the systematic research on the relationship between the farmland fragmentation and rural collective actions in China.

2.1.1.1 Concept, Characteristics and Status Quo of Farmland Fragmentation

(1) Concept and definition of farmland fragmentation

In 1950, Binns (1950) wrote the first book that comprehensively reflected the governance of fragmented agricultural land, and defined the concept of farmland fragmentation for the first time. Binns defined farmland fragmentation as a structure of spatially disconnected use of farmland under the common impact of natural conditions and human activities. Then, many scholars also provided their definitions of this concept. Among them, the definition made by King and Burton (1982) has been generally recognized. They believed that the farmland fragmentation should be mainly defined by two conditions. One is that the farmland has been divided into multiple plots, and the other is that each plot is too small in area to have proper development. Later, many scholars defined and classified the concept of farmland fragmentation for different research purposes. For example, Sabates-Wheeler (2002) defined from four aspects: physical fragmentation, social fragmentation, activity fragmentation, and ownership fragmentation; Dijk (2003) defined from four perspectives: ownership fragmentation, use fragmentation, internal fragmentation, and the discrepancy between ownership and use.

Many Chinese scholars have also defined the concept of farmland fragmentation. For example, Tai Xiaoli pointed out that farmland fragmentation refers to the phenomenon that peasant households contracting farmland have multiple plots of farmland each covering a small operating area, and it is a land use pattern contrary to the large-scale operation of farmland. That is, due to human or natural factors, the farmland is divided into scattered, dispersed patches of different sizes, showing a disorderly scattered state (Tai, 2016).

From a different perspective, the interpretation of farmland fragmentation will also be different. From an institutional perspective, farmland fragmentation is based on the principle of fairness. Under the differentiated distribution institutional rules, the entire farmland is allocated to different peasant households, so that a single peasant household owns multiple plots differed in soil fertility, area, shape and distance (Wang, 2016). From a spatial perspective, farmland fragmentation refers to a form of farmland structure in which the farmland of a single peasant household is divided into multiple plots disconnected to each other in space. From the perspective of ownership, farmland fragmentation means that a single peasant household has the right to contract and manage multiple plots of farmland that are disconnected to each other in geographical space. From the perspective of natural environment, farmland fragmentation is caused by various factors including topography. It is difficult to integrate, concentrate or operate the farmland on a large scale, and the use of farmland falls in a scattered, dispersed and disordered state. It can be found that the interpretation of farmland fragmentation is made quite differently due to different emphasis from various perspectives.

Although the concept of farmland fragmentation has not yet been unified, it mainly consists of four parts. First, natural fragmentation. It mainly refers to the fact that farmland is fragmented spatially due to natural factors, such as natural disasters and topographic structure (Sabates-Wheeler, 2002). Second, ownership fragmentation. During the distribution of farmland ownership, the ownership of farmland or the right to contract management is held by multiple people, due to the principle of fairness and the implementation of a differentiated distribution system which requires farmland to be distributed by matching fertility and distance, so that each owner only holds many small plots of farmland (Dijk, 2003). Third, agricultural operation fragmentation. It is actually a subdivision of ownership fragmentation, with more emphasis on the division between land owners and users (Niroula, 2005). For example, land owners divide and lease large plots of farmland to different peasant households for agricultural production. Fourth, non-agricultural operation fragmentation. Opposite to the concept of agricultural operation fragmentation, it mainly emphasizes the land use structure of spatially separated farmland caused by non-agricultural operations, such as road and canal construction (Hartvigsen, 2014).

Based on the above characteristics of farmland fragmentation, this book provides the following definition of farmland fragmentation, that is, farmland fragmentation refers to the phenomenon that contracted peasant households own multiple plots of farmland that are geographically disconnected, and each plot fails to create economy of scale. But this phenomenon can be changed through the exchange of farmland.

(2) Characteristics of farmland fragmentation

The above four dimensions comprehensively interpret the concept of

CHAPTER 2
LITERATURE REVIEW AND THEORETICAL BASIS

farmland fragmentation. Due to different perspectives of definitions, the phenomena of farmland fragmentation may not be able to fully reflect the above four dimensions. In specific practices, scholars agree that farmland fragmentation must present three characteristics (Wang, 2008).

Firstly, it is geographically divided. A peasant household has multiple plots of farmland that are not adjacent to each other, that is, the farmland of the family is disconnected with each other in spatial and geographical location. Specifically, the number of plots can be directly used to reflect the degree of farmland fragmentation. The more the number of plots, the higher the degree of farmland fragmentation. In addition, the spatial and geographic non-adjacency of farmland can also be reflected by distance, that is, the distance between plots or between the land and the residence of peasant household can be used to explain the degree of farmland fragmentation (Binns, 1950).

Secondly, the plots are in small areas. That is to say, a plot of farmland only covers a small area. Due to the different physical and geographical conditions in different regions, it is impossible to use specific values to measure the size of a plot of farmland, but it can be measured by whether the plot create economy of scale, that is, with the gradual increase in the area of a single plot, whether the role played by other factors of production on the plot is also gradually improving, whether the overall efficiency can be gradually raised, whether the input cost of a single plot is decreasing or whether the output increases. If the plot cannot create economy of scale, it means the average plot area is rather small, and there is the phenomenon of farmland fragmentation.

Thirdly, the ownership of plots should be exchangeable. That is to say, though manifested by the fragmentation of land plots, farmland fragmentation is by nature the ownership fragmentation. The phenomenon of farmland fragmentation can be solved through plot exchange (Tian et al., 2015), that is, a single plot can be exchanged by transferring the contract management rights of farmland and other means, so as to realize the consolidation of plots and form a certain operation scale (King, 1982).

(3) Status quo of farmland fragmentation

1) Farmland fragmentation all over the world

The phenomenon of farmland fragmentation can be found widespread in many countries around the world. From a global perspective, the areas with the most severe degree of farmland fragmentation mainly include Central and Eastern Europe, India and China (Sklenicka, 2014). For example, the average area of Polish farms is about 8 ha, and about 18% of the farms have more than 6 land plots (Janus, 2017); the average farmland area among 1.78 million farms in Bulgaria recorded 2.4 ha, and each farm owns 4.5 plots on average, with each plot covering an area of 0.5 ha (Di, 2010); farms in Turkey has 4.1 plots of farmland on average, with each plot covering an area of about 1.5 ha (Yucer, 2016). In addition, some scholars, according to their own research purposes, obtained data from some areas through field investigations and discussed about farmland fragmentation. For example, Latruffe (2014) pointed out that each farm in Brittany in France had 14 plots of farmland on average, with each plot covering 4.35 ha; Abubakari (2016) mentioned that, in the northern ecological zone of Cannes, 43% of the peasant households owned 3 plots of farmland, 27% of the peasant households owned 2 plots of farmland, 23% of the peasant households owned 4 plots of farmland, and 7% of the peasant households owned more than 5 plots of farmland; Abdollahzadeh (2012) argued that in the Khur Plain of Esfahan Province, Iran, each farm had an average of 5.29 plots, each covering 0.07 ha.

From the historical changes in farmland fragmentation, the degree of farmland fragmentation in some countries gradually slows a down trend. For example, in the early 1990s, the average area of farms in Albania was 1.05 *ha*, which were divided into 2 to 5 plots. By 2011, the average area of farms increased to 1.26 *ha*, with an average of 4.7 plots per farm (Hartvigsen, 2014). In Japan, from 1985 to 2005, the average area of rice fields increased from 0.74 *ha* to 0.9 *ha*. The average area of paddy fields less than 0.5 *ha* was 0.35 *ha*, and each peasant household had an average of 2.2 plots; the average area of paddy fields within 0.5–1 *ha* range was 0.73 *ha*, and each peasant household had an average of 3.03 plots; the average area of paddy fields within 1–2 *ha* range was 1.43 *ha*, and each peasant household had an average of 4 plots; the average area of paddy fields within 4–8 *ha* range was 5.6 *ha*, and each peasant household had an average of 6.11 plots; the average area of paddy fields over 8 *ha* was 16.2 *ha*, and each peasant household had an average of 9.29 plots (Kawasaki, 2010).

However, the phenomenon of farmland fragmentation in some other countries gradually aggravates. Sklenicka (2016) studied the historical changes and future trends of the farmland ownership fragmentation in the Czech Republic over the past 230 years. The study showed that from 1785 to 2015, the average area of 102,984 plots in the Czech Republic decreased from 1.08 *ha* to 0.64 *ha*, an average annual decrease of 0.26%; the number of land holders per 100 *ha* increased from 17.5 to 79.66, an average annual growth of 0.61%. The average number of plots of land holders in Sri Lanka increased from 2.5 in 1962 to 3.2 in 1982 (Thiesenhusen, 1990); the average area of Indian farms decreased from 1.57 *ha* in 1990 to 1.33 *ha* in 2000 (Manjunatha, 2013); the average area of a single plot in Bangladesh decreased from 0.16 *ha* in 1977 to 0.12 *ha* in 1996 (Rahman, 2008).

2) Farmland fragmentation in China

On the whole, the current degree of farmland fragmentation in China is more serious than that in other countries. Due to the complexity and diversity of the measurement of farmland fragmentation, there is still no official data on farmland fragmentation. From the perspective of farmland per capita, from 1983 to 2012, the farmland per capita operated by rural households in China always changed within the range of 1.93 *mu* to 2.34 *mu*①, and the average farmland operated by each household remained at about 8 *mu*. From the perspective of the number of plots, according to the "Hundred Villages and Thousand Households" survey conducted by China Institute for Rural Studies, Tsinghua University, the average farmland held by rural households in China keeps basically between 4 and 5 plots. ②

Because of China's vastness in territory, the degree of farmland fragmentation varies in different provinces. For example, Wu Mingfeng's survey of 11 provinces across China showed that the average number of plots per household was about 5.28, each covering 0.13 *ha* (Wu, 2017); Ji Yueqing's survey of Xuancheng City of Anhui Province and Taixing City of Jiangsu Province showed that the average number of plots in the two places was 5.12 and 2.99 respectively, each covering 0.08 and 0.2 *ha* (Ji, 2016); Guo Guancheng's survey of Xuzhou and Yancheng in Jiangsu showed that the average number of plots per household was 3.51 and 4.1, each covering about 0.1 *ha* (Guo, 2016). On the whole, the degree of farmland fragmentation in China is relatively serious, especially the small area of each plot, making it difficult to form a scale effect compared with other countries in the world.

2.1.1.2 Causes of Farmland Fragmentation

There are many reasons underlying farmland fragmentation. King

---

① Based on the database of National Bureau of Statistics of China at http://data.stats.gov.cn.

② Calculated based on the survey results of China Institute for Rural Studies, Tsinghua University, in 2014 and 2017.

CHAPTER 2
LITERATURE REVIEW AND THEORETICAL BASIS

(1982) believed that the reasons for farmland fragmentation in Europe mainly included institutional, political, historical, social and other factors, such as inheritance law, centralization system, land transaction costs, urban development planning, and personal valuation; Di (2010) further pointed out after studying that the inheritance system and distribution system in Europe were the main reasons for farmland fragmentation. From a historical point of view, the equivalence system of family properties in China is the most important reason for the long-term farmland fragmentation, and the recent implementation of the Household Contract Responsibility System is the main reason for the current phenomenon of farmland fragmentation (Xie et al., 2017). In addition, objective factors of China, such as backward market transaction mechanism and huge population, are also the main reasons for this phenomenon. Generally speaking, the occurrence of farmland fragmentation is mainly attributed to the following four factors.

(1) Natural factors

Natural factors are the primary factor leading to farmland fragmentation, and also the most important constraint against the solution of farmland fragmentation. Natural factors mainly include geological structure, land quality, natural disasters and many other aspects. Among them, the spatial and geographic division of farmland caused by topographic structure is the main reason for farmland fragmentation (Abubakari, 2016). In addition, some scholars also further argued that the farmland in mountainous or hilly areas was more likely to show a spatially divided land structure than plain areas, due to the constraint of topographical conditions (Yucer, 2016).

(2) System factors

Among the system factors, the distribution system exerts particularly important impact. Under normal circumstances, the distribution of farmland ownership needs to be based on the principle of fairness involving land quality, distance, risk aversion, and crop diversity, which caused a single peasant household to hold several plots disconnected geographically. The distribution system formed by multiple external forces has led to the occurrence of farmland fragmentation (Di, 2010). In addition, the transaction system is another factor that causes farmland fragmentation. Kawasaki (2010) explained that in order to avoid market risks, farmland holders did not tend to transfer large plots at once. Instead, they divided them into multiple plots and gradually transfer to others, resulting in a small area of each plot. For example, the average area of each plot for transaction in Japan is only 0.2 to 0.3 *ha*. What's more, due to the imperfect farmland transaction mechanism in some areas, it is difficult to guarantee the interests of farmland transfer. Therefore, they tend to choose trustworthy holders like relatives and friends for farmland transfer, so that distance and area of plots become secondary factors, and farmland fragmentation has been further intensified.

(3) Social and cultural factors

Among the social and cultural factors, the custom of property inheritance is the primary reason leading to farmland fragmentation. Under the constraints of local customs, farmland is usually evenly allocated to each inheritor. As the number of family descendants increase, the limited area of farmland resources is allocated to more and more heirs, therefore, the total area of farmland held by the heirs keeps decreasing, and farmland fragmentation becomes increasingly severe (Niroula, 2005). At present, the customs of property inheritance in many countries around the world have led to the phenomenon of farmland fragmentation, such as Ghana (Abubakari, 2016), Japan (Kawasaki, 2010), the Czech Republic (Sklenicka, 2016), and Turkey (Yucer, 2016). In addition to customs of inheritance, donation as one of the social and cultural factors is another reason for farmland fragmentation. Since farmland is an important property, part of the land is often given as a dowry or gift between relatives and friends. The ownership of farmland is transferred to other families, causing the land ownership to be increasingly fragmented (Abdollahzadeh, 2012).

(4) Economical factors

The degree of economic development is another important factor leading to the phenomenon of farmland fragmentation. It is believed by some scholars that the income heterogeneity among peasant households has aggravated farmland fragmentation. Under normal conditions, after a peasant household holding large plots of farmland accumulates a certain amount of capital in the agricultural industry, it will shift to engage in a more profitable non-agricultural industry and thus transfer the farmland (Qiu, 2015). On the one hand, the holder divides large plots of farmland into several small plots for gradual transfer, in order to avoid market risks. On the other hand, the peasant household with small plots often has poor economic conditions and is not capable of buying large plots. Therefore, the land holder can only transfer the plots they own to multiple peasant households, which leads to farmland fragmentation (Di, 2010). What's more, the development of urbanization has also led to farmland fragmentation to a certain extent. With the expansion of cities, more farmland was converted to non-agricultural land. The total area of farmland decreased, while the population increased sharply, causing the area of each plot accordingly decreased.

2.1.1.3 Evaluation Method of Farmland Fragmentation

Since scholars have different definitions of the concept of farmland fragmentation, the evaluation indexes of farmland fragmentation are also varied. Judging from the extant literatures, the evaluation methods for farmland fragmentation are mainly summarized into two, namely, economic methods and ecological methods.

(1) Economic methods

The economic methods mainly evaluates the degree of farmland fragmentation from the perspective of whether a single plot create economy of scale, usually with quantifiable indexes, such as the area of each plot, the total number of plots, and the distance of the plots each other.

The total number of plots interprets the characteristics of the degree of spatial and geographic disconnection of plots in farmland fragmentation. Assuming that different peasant households have the same total area of contracted plots, the larger the number of plots contracted by households, the smaller the area of each plot, resulting in more serious farmland fragmentation. According to the extant research, most scholars use the quantitative index of the total number of plots held by a peasant household as an important index to evaluate the degree of farmland fragmentation. That is, they obtained data of land contracted by peasant households by means of social support questionnaire surveys in rural areas, to evaluate their degrees of farmland fragmentation. For example, Ji Yueqing et al. (2016), Zhang Yinjunjie et al. (2008), Qin Lijian (2011), Liu Qijun (2011), Xu Qing (2008), Li Gongkui (2006) used this index as the only index to evaluate the degree of farmland fragmentation. However, although this index may interpret the geographical disconnection, there may be strong economy of scale in many plots. Therefore, some scholars also included "the number of plots covering less than one *mu*" owned by a peasant household as an evaluation index for farmland fragmentation (Wu, 2017).

The area of each plot directly reveals the degree of farmland fragmentation from the perspective of economy of scale, that is, the smaller the area of each plot, the greater the possibility of creating economy of scale on the plot, and the higher the degree of farmland fragmentation. Lian Xuejun (2013) and Wu Yang (2008) used the area of each plot as an evaluation index to measure farmland fragmentation.

Besides the area index, distance is also an important index to quantify the degree of farmland fragmentation. The distance index can well explain the spatial and geographic distance of a single plot, which determines the input of a peasant household in production factors, such as time cost, labor cost, transportation cost, etc. A far distance indicates that the peasant household invests more in the production process. The advantage of economy of scale lies in the law that as the output increases, the long-term average total cost will reduce. But the increase in the long-term average total cost caused by the far distance of plots is much greater than the advantage of cost reduction brought by the economy of scale. Therefore, the longer the distance, the higher the degree of farmland fragmentation. According to the extant literatures, scholars mainly use two specific quantitative indexes, the average distance between multiple plots and the average distance between all single plots and the residence, to evaluate the degree of farmland fragmentation. For example, Wen Gaohui (2016) involved the average distance between the plots and the residence as an important index for the evaluation of farmland fragmentation in addition to the number of plots and the average area of each plot.

The above three indexes all quantify the degree of farmland fragmentation from different perspectives. However, if one of the indexes is used alone to measure the degree of farmland fragmentation, it will fail to fully reflect all the characteristics of farmland fragmentation. The application of all the three evaluation indexes requires reasonable weights. Therefore, how to build a comprehensive evaluation index that can fully cover the three factors of the number of plots, the area and the distance is particularly important for the measurement of the characteristics of farmland fragmentation. To this end, King constructed indexes S, J and I to evaluate the degree of farmland fragmentation. The specific calculation formula is as follows:

$$S = 1 - \frac{\sqrt{\sum_{i=1}^{n} \alpha_i^2}}{(\sum_{i=1}^{n} \alpha_i)^2} \qquad (2-1)$$

$$J = \frac{\sqrt{\sum_{i=1}^{n} \alpha_i}}{\sum_{i=1}^{n} \alpha_i} \qquad (2-2)$$

$$I = n / \sum_{i=1}^{n} \alpha_i \times \frac{1}{100} \times \sum \omega \qquad (2-3)$$

In the above formula, i, α, n, and w respectively represent the i-th plot, plot area, total plot number, distance between plots, or distance between plots and residences. The value range of S and J is between 0 and 1. The larger the S index, or the smaller the J index, or the larger the I index, the more serious the degree of farmland fragmentation. However, some scholars pointed out that the above indexes also have certain limitations when evaluating farmland fragmentation. That is, it is difficult to reflect the importance of plot number and area in the impact of farmland fragmentation in indexes S and J. Besides, it is also difficult for index I to measure the spatial distribution of plots (Shen, 2012). Thus, some scholars made improvements on the evaluation indexes of farmland fragmentation based on the above three indexes, for example introducing a spatial adjacency matrix to further deepen the quantitative evaluation method of farmland fragmentation.

(2) Ecological methods

Current scholars mainly use landscape ecology methods to evaluate the degree of farmland fragmentation, that is, use patches of farmland as the basic unit of landscape composition, and use the landscape pattern index to reflect the information of farmlandscape pattern, so as to reflect specific characteristics such as the structural composition and the spatial configuration of farmland within a certain area. The extant literatures mainly establish diversified landscape index for evaluating farmland fragmentation focusing on the average area, shape and distribution of the farmland (Demetriou, 2013; Latruffe, 2014). Specifically, there are six indexes, i. e. average patch area, patch density, boundary density, area-weighted shape index, area-weighted fractal dimension, and patch aggregation index (Wang, 2016; Tai, 2016; Wang, 2014).

The average patch area is the most basic index for the evaluation of the degree of farmland fragmentation in the landscape index. It refers to the ratio of the total area of farmland to the number of plots within a certain area. The smaller the average area of the patch, the severer the degree of farmland fragmentation (Huang, 2015; Li, 2012). The explanation of patch density and average patch area is completely opposite. It refers to the ratio of the number of plots to the total area of farmland within a certain area, that is, the number of patches per unit area. Under normal circumstances, the larger the patch density, the severer the degree of farmland fragmentation (Huang, 2015). Boundary density refers to the length of the boundary per unit area of the landscape formed by the use of farmland. It is usually presented by the ratio of the total perimeter of the plots to the total area of farmland within a certain area. The larger the boundary density, the severer the farmland fragmentation (Wang, 2014). The area-weighted shape index reflects the problem of plot shape in the degree of farmland fragmentation. It uses square as the basic measurement standard. The specific calculation is like the ratio of the perimeter of each block of a kind within the area should be multiplied by their respective area weight before calculating the sum. The area-weighted shape index value is equal to 1, indicating that the shape of the farmland block is a simple square. The larger the value, the more complex the shape of the block, that is, the severer the farmland fragmentation (Huang,

2015). The area-weighted fractal dimension is a further improvement of the area-weighted shape index. It also reflects the problem of farmland shape in the degree of farmland fragmentation. Its value range is between 1 and 2. The value of 1 means the block is in the simplest square or circle. The value of 2 means the block is the most complicated type. That is, the closer the index is to 2, the severer the farmland fragmentation. Patch aggregation index reflects the degree of aggregation of used patches in the landscape, that is, the dispersion of landscape elements in the farmlandscape. If the value of the patch aggregation index is low, it means that the farmland is composed of many small patches, that is, a severer degree of farmland fragmentation. The above six indexes are often used in extant researches because they can more comprehensively and reasonably reflect the various characteristics of farmland fragmentation. Moreover, some scholars use other landscape pattern indexes to further quantify the degree of farmland fragmentation, such as the spatial dispersion, dominance, and uniformity of plots (Sun, 2015).

However, although the above indexes can quantify the degree of farmland fragmentation to a certain extent, they only differ in their focus, without simply reflecting one aspect of the characteristics of farmland fragmentation. Therefore, the indexes feature a strong colinearity. If these indexes are used to measure farmland fragmentation at the same time, there will be a big difference between the measured degree of farmland fragmentation and the actual situation. Therefore, on the basis of calculating the specific values of the above indexes, some scholars further introduced the method of principal component analysis to refine the landscape pattern index, construct a principal component analysis with a relatively small degree of colinearity, and use it as a specific index for the evaluation of the degree of farmland fragmentation. The research by Guo Hui (2016), Wang Daojun (2014) and Tai Xiaoli (2016) are the example of the application of this method.

(3) Other methods

The above two methods are currently the main methods for measuring farmland fragmentation. In addition, scholars also use other methods to evaluate farmland fragmentation. For example, Lyu Zhenyu (2014) held that the evaluation index system for the degree of farmland fragmentation should cover the spatial and social attributes of the plot. Therefore, multiple indexes may be selected to build a three-level index system, and a specific evaluation system can be built by using the analytic hierarchy process. It is further described in their research that according to the importance of the indexes, the specific indexes are respectively the average area of the plot, the consistency of the length of the plots, the average number of plots of the peasant household, the square field rate, the consistency of the row direction of the plots, and the gap between the peasant household's actual farmland operation scale and moderate scale, road accessibility of the plot, etc. Shen Chenhua (2012) established an index calculation method for measuring the degree of farmland fragmentation based on whether to consider the factor of interaction between spatially adjacent plots.

On the whole, the above methods for evaluating farmland fragmentation have certain advantages and disadvantages. The economic methods focus on reflecting the degree of farmland fragmentation of a certain individual, which can fully show the situation of ownership fragmentation. However, because the data applied usually come from questionnaire surveys, it is difficult to fully reflect the farmland fragmentation of all farmland within a certain area due to the limitations of the survey. The data applied in the ecological method is mainly obtained from geographic databases, such as the remote sensing image, which can fully reflect the actual situation of farmland fragmentation in a specific area. But it mainly describes farmland fragmentation from the perspective of physical geography, and cannot reflect the phenomenon of farmland fragmentation caused by ownership factors. Thus, there are also certain defects (Wen, 2016).

2.1.1.4 Impact of Farmland Fragmentation

(1) Impact based on the perspective of the characteristics of land fragmentation

1) Impact of the number of plots

Extant researches have shown that the number of plots in land fragmentation exerts two kinds of effects. Generally, the negative impact of the number of plots is mainly manifested in three aspects: Firstly, a greater number of plots increase the cost of capital investment. Since every single plot requires such infrastructure as irrigation and roads, peasant households must invest in the infrastructure construction, causing more capital investment than those with a small number of plots (Janus, 2017; Tan, 2006). Secondly, a greater number of plots increase organization and management costs. Peasant households holding multiple plots should allocate their limited capital, labor and other means of production among different plots in a proper way, which results in higher organization and management costs in the agricultural production process (Sklenicka, 2016). Thirdly, a greater number of plots increase the laborers' time input. Due to the great number of plots, the farmers need to travel to and fro between different plots to manage their agricultural production, which greatly increases the laborer's time input. Meanwhile, however, current researches show that the number of plots also exerts a certain positive effect. On the one hand, farmers can make the best choice of crops according to the fertility conditions of each plot (Farley, 2012). Farmers holding more plots tend to carry out diversified planting, and can make reasonable arrangements according to the actual conditions of each plot, to realize optimal use of each plot, maximize crop yields, and thus improve their agricultural income. In addition, diversified planting can also bring in higher landscape value, which in turn may increase the non-agricultural income for farmers through rural tourism and other means. On the other hand, multiple plots reduce the risk of agricultural production. Since multiple plots are disconnected to each other in space, losses caused by natural disasters can be effectively avoided, so as to reduce the natural risks faced by farmers in agricultural production (Farley, 2012; Looga, 2018).

2) Impact of plot area

Most researches indicate that the smaller the plot area, the greater the negative impact (Guo, 2016). Firstly, the smaller plot area reduces the economy of scale. Under normal circumstances, the smaller the plot area, the higher the production cost per unit of agricultural products, and the lower the marginal output elasticity of input factors in the agricultural production process, which will greatly weaken the economy of scale for agricultural production. Secondly, the smaller plot area leads to the abandonment of arable land. In the case of holding multiple plots, some farmers may choose to discard some of the smaller ones due to their lower economy of scale, thus the efficiency of land use will seriously decrease (Kjelland, 2007; Sheng, 2014). Thirdly, extant researches demonstrate that the smaller plots are usually lower in market transaction prices. Nevertheless, some argued that smaller plots also had certain positive impacts. If the peasant household holds a large number of plots, each in small areas, it will usually exchange labor with others to achieve labor input in agricultural production activities, thereby reducing the cost of hiring labor to a certain extent.

3) Impact of distance

Generally speaking, most scholars hold that distance has more negative impacts, especially in the aspect of agricultural production cost. Firstly, distance leads to higher transportation costs. A long distance between plots will increase the cost for the transportation of agricultural means of production and agricultural products (Jacoby, 2002). Secondly, distance increases the labor input cost. Because farmers must travel to and fro between different plots to carry out agricultural production, compared with peasant households travelling shorter distances, farmers with plots far apart have to invest more time, which also increases the difficulty of proper family labor distribution, and causes a decrease in labor productivity (Dirimanova, 2010). Thirdly, distance also reduces the efficiency of land use. Under normal circumstances, farmers are less willing to invest in plots located far away, so their index of multiple cropping may decline, resulting in a severe phenomenon of abandonment of specific plots and lower efficiency of land use (Deininger, 2012). In addition, some scholars have argued that distance also leads to more additional losses. For example, a longer distance causes more losses in the transportation of irrigation water, and meanwhile increases the difficulty of irrigation management. Therefore, longer distance between plots results in lower efficiency of irrigation. A few scholars have also argued that distance may as well exert certain positive effects in agricultural production. Just as different crops have different requirements for management and protection inputs, distance can also promote the rational allocation of crops in different plots. For example, they can plant crops with lower management and protection requirements in plots far away from their residence, and plant crops with higher management and protection requirements in plots close to their residence, thereby realizing a proper distribution of labor (Thiesenhusen, 1990).

4) Impact of plot shape

Extant researches have reached a consensus that the shape of plots has more negative impacts on agricultural production. Firstly, the irregularly shaped plots restrict the functioning of agricultural machinery. Due to the irregular shape of the plot, agricultural machinery and equipment cannot be used on a large scale, so that it becomes difficult to exert technical advantages in the agricultural production process. For labor-intensive agriculture, agricultural production can only be carried out by men rather than machinery, causing a far reduced production efficiency of farmers (Lam, 2018). Secondly, the irregularly shaped plots feature poor infrastructure accessibility. Due to the extensive border areas of irregularly shaped plots, the agricultural infrastructure, including those for irrigation, cannot reach all the border areas. Thus, agricultural production fails to meet basic needs, which further lowers the level of crop production (Lu, 2016). What's more, some scholars believed that irregularly shaped plots would cause more losses. More border areas lead to a large amount of losses in the agricultural production process, especially more losses in the process of crop harvesting, which significantly reduces the level of agricultural output.

(2) Impact based on the perspective of agricultural production

1) Farmland fragmentation and agricultural production efficiency

Since the publication of Schultz's *Transforming Traditional Agriculture* in 1964, researches on the relationship between farmland fragmentation and agricultural production efficiency have gradually increased (Niroula, 2004). Most researches indicate that farmland fragmentation reduces the agricultural production efficiency (Hartvigsen, 2014; Rahman, 2008), which is mainly due to the difficulty in achieving operation scale in agriculture, in realizing effective allocation of means of production, and in using agricultural machinery. Thereby, some scholars carried quantitative analysis on the relationship between farmland fragmentation and agricultural production efficiency. For example, Rahman (2008) found that a 1% increase in the degree of farmland fragmentation in Barisal region of Bangladesh leads to a 0.03%

reduction in rice production efficiency.

However, scholars represented by Schultz believed that although the number of plots and other factors had reduced the agricultural production efficiency to a certain extent, the small area of farmland was conducive to the improvement of agricultural production efficiency (Schultz, 1964). On the one hand, the farmland with a small area facilitates the farmer to conduct effective management and fully use agricultural input elements, so the agricultural production efficiency will be improved (Manjunatha, 2013); on the other hand, since agriculture is a labor-intensive industry and requires a lot of labor input, the farmer with smaller plots may resort to mutual resource borrowing and other methods for agricultural production, such as mutual labor borrowing, which helps save the cost of hiring labor. Therefore, the farmer with smaller plots is able to increase agricultural input-output ratio by reducing costs. However, with the continuous innovation and application of agricultural machinery, agricultural technology, and agricultural production and operation methods, the above viewpoints are gradually weakening (Niroula, 2004).

Judging from the extant literatures, most Chinese scholars believed that farmland fragmentation had a certain negative effect on agricultural production efficiency. The main reason is that farmland fragmentation leads to waste of agricultural means of production, more intensive labor input, and higher costs for supervision and management, and that farmland fragmentation causes an increase in the proportion of the land area of infrastructure such as irrigation ditches and shelter forests. Thus, under both direct and indirect impact, farmland fragmentation will eventually exert strong negative impacts on agricultural production efficiency. For example, Li Yinqiu (2011) studied the relationship between farmland fragmentation and input-output efficiency in the process of rice production, found out that the phenomenon of farmland fragmentation greatly reduces the input-output efficiency of rice production, and especially exerts severe impacts on the efficiency per unit area of paddy field. Specifically, the most important reason is that farmland fragmentation has greatly increased farmers' management cost for rice production. In addition, it restricts the application of some new technologies and equipment, leading to higher corresponding application costs. The multi-dimensional effects of farmland fragmentation in the agricultural production process together lead to a decline in agricultural production efficiency.

The impact of farmland fragmentation on agricultural production efficiency comes from multiple dimensions. In addition to its impact on the overall agricultural production efficiency, current scholars also probed into the relationship between farmland fragmentation, technical efficiency, and land use efficiency in the agricultural production process, before reaching a consensus that farmland fragmentation also has a negative impact on agricultural production efficiency. In terms of the technical efficiency of agricultural production, Li Xin (2012) argued that farmland fragmentation had a serious inhibitory effect on the improvement of the technical efficiency of agricultural production. Huang Zuhui (2014) also held that the higher the degree of farmland fragmentation of rice farmers, the lower the technical efficiency. The main reason is that the small area, irregular shape, and discrete distribution of single plots limit the scale utilization of agricultural machinery and equipment and hinder the implementation of advanced agricultural technology. Farmland fragmentation makes it difficult to exploit the technical advantages in the agricultural production process, and ultimately leads to lower technical efficiency of agricultural production (Li, 2012). In addition to the technical limitation of farmland fragmentation, which reduces the technical efficiency of agricultural production, farmland fragmentation also indirectly affects the technical efficiency of agricultural production through changes in other input factors. The severer the degree of farmland fragmentation, the larger number of plots and the smaller area of each plot in the peasant households, as well as the more unreasonable the allocation of labor and capital investment in each plot (Zhang, 2012).

Apart from the above researches on the technical efficiency of agricultural production, some also probed into the impact of farmland fragmentation on land use efficiency, and concluded that farmland fragmentation increased the difficulty of farming, thus not conducive to farmers' investment in land, resulting in lower multiple cropping index of farmers and hindering the improvement of land use efficiency on average (Liu, 2008). In addition, some also analyzed other aspects of efficiency in the agricultural production process. For example, Zu Jian (2016) studied the impact of farmland fragmentation on farming efficiency, distance efficiency and irrigation efficiency in mountainous and hilly areas in China, which pointed out that the greater the degree of farmland fragmentation, the lower the farming efficiency, distance efficiency and irrigation efficiency would be.

However, some argued in their researches that the impact of farmland fragmentation on agricultural production efficiency was actually not obvious. Because in China, the agricultural industry remains a labor-intensive industry, and smallholders are still the mainstay of China's agricultural operations. Considering China's large population, farmland fragmentation is conducive to the allocation of rural surplus labor in different plots, that is, helping absorb the surplus rural labor. Therefore, farmland fragmentation does not necessarily lead to the reduction of agricultural production efficiency (Lian, 2014). In the research by Bai Zhiyuan (2015) on the relationship between farmland fragmentation and land use efficiency, he did not find any obvious correlation between the two. Some scholars also pointed out that farmland fragmentation was conducive to the improvement of land use efficiency, mainly because the soil could not lie fallow sufficiently under continuous cultivation. Continuous cultivation on a piece of arable land may easily cause decline in soil quality. On the contrary, most farmers with fragmented land usually rotate farming on different plots, so farmland fragmentation is conducive to the improvement of land use efficiency (Ram, 1999).

2) Farmland fragmentation and agricultural production costs

Extant researches agree that farmland fragmentation has increased the agricultural production costs to a large extent (Vidal-Macua, 2018; Latruffe, 2014; Abdollahzadeh, 2012; Sklenicka, 2014; Latruffe, 2014). Generally, the extant researches all hold that farmland fragmentation increase the total input in the agricultural production process, but in terms of individual costs, their researches present differentiated viewpoints. Lu Hua (2015; 2016; 2017) argued that farmland fragmentation increased the total cost of crop per unit output in the agricultural production process, but it has a certain differentiation in the impact direction for individual costs. Specifically, farmland fragmentation has significant positive effects on labor costs and fertilizer costs, that is, the higher the degree of farmland fragmentation, the higher the costs of these two items. However, the impact of farmland fragmentation on seed costs, mechanical service costs, and animal input costs turned out to be negative, that is, the greater the degree of farmland fragmentation, the lower the costs of these items. The reason is that as the degree of farmland fragmentation intensifies, farmers may adjust agricultural production input factors between different plots. Because there is usually labor surplus in rural areas in China, farmland fragmentation may often increase the input of labor force. Farmers holding fragmented land tend to choose diversified planting, so the seed input cost will be reduced. Besides, due to farmland fragmentation, it is difficult to apply mechanical services and animal power, which will reduce farmers' capital input costs in mechanical services and animal power. Farmland fragmentation has different impacts on different individual costs in the agricultural production process, which also further illustrates that farmland fragmentation may encourage farmers to allocate production factors more flexibly and change the input of production factors among different plots. On the whole, however, the labor input is to offset capital expenditures including machinery input, that is, farmland fragmentation may prompt farmers to choose labor-intensive production methods and reduce the use of capital-intensive agricultural production methods, for example agricultural machinery.

Wu Yang (2008) argued that farmland fragmentation increased the productive input of farmers, because farmland fragmentation makes farmers choose diversified crop planting, which leads to increased investment. In addition, farmland fragmentation reduces the scale effect, that is, the amount of pesticides and fertilizers invested per unit area of land will increase, resulting in higher costs. Yang Zhaoxi (2017) believed that farmland fragmentation would increase both labor costs and capital input costs. In terms of labor costs, the higher the degree of farmland fragmentation, the more likely will farmers replace agricultural machinery input with labor input. In terms of capital input costs, farmland fragmentation increases the operation time and loss of agricultural machinery for land preparation, pumping, etc., thereby increasing agriculture mechanical input.

3) Farmland fragmentation and agricultural production income

Since farmland fragmentation affects agricultural production costs through many ways, agricultural production income is also affected to a certain extent. Most scholars believed that farmland fragmentation reduces agricultural production income. Latruffe (2014) studied the impact of farmland fragmentation on farm performance in Brittany, France, and found out that the number, area, distance and other characteristics of farmland fragmentation had adverse effects on farm performance. Di (2010) made researches on the relationship between farmland fragmentation and agricultural production efficiency in Plovdiv region, Bulgaria, and also confirmed the above view, that is, the severer the farmland fragmentation, the lower the agricultural production income. Rahman (2008) held that the most important negative impact of farmland fragmentation on agricultural production income is the reduction of crop yields, and demonstrated through empirical research that a 1% increase in the degree of farmland fragmentation in Barisal region, Bangladesh led to a 0.05% reduction in paddy yield.

Some scholars also compared and analyzed the costs and incomes of farmland fragmentation. For example, Kawasaki (2010) studied the impact of farmland fragmentation on the production costs and incomes of farms that mainly produce paddy in Japan. It is indicated that the production income generated by the scale of arable land is far from enough to offset the production cost generated by the number of plots, and although farmland fragmentation may avoid agricultural production risks to a certain extent, its monetary value is far lower than the negative impact of farmland fragmentation on production costs. Thus, generally speaking, the negative value of farmland fragmentation in term of costs is much higher than the income it brings.

At present, Chinese scholars also discuss the impact of farmland fragmentation on farmers' income. Most of them believed that farmland fragmentation had a positive effect on the improvement of farmers' agricultural and non-agricultural incomes. The reasons can be summarized as follows.

First of all, farmland fragmentation encourages farmers to diversify planting, which resolves the natural risks and market risks of agricultural production, and is conducive to the increase of agricultural income. Farmland fragmentation may drive farmers to choose different types of crops on different plots, thus increasing the types of crops, and effectively resisting the negative impacts of natural disasters such as pests, diseases and geological disasters. Especially in areas with frequent occurrence of natural disasters, farmland fragmentation improves the ability of farmers to respond to natural risks. In addition, through diversified planting of crops, farmers may effectively avoid market risks caused by sharp fluctuations in the price of single variety crops in the market. Especially for high value-added agricultural products such as vegetables, the diversification of planting helps farmers endure less market risks. In addition, diversified planting may encourage farmers to rationally arrange their labor hours according to the fruit seasons of different crops, thus optimizing agricultural production time and increasing income levels (Li, 2006). What's more, diversification of planting helps different types of crops to match different soil types and climatic conditions, and also helps farmers to use different fertilizers to meet the needs of various crops, improve soil fertility, and thereby increase agricultural output and increase agricultural income (Lu, 2015). Therefore, the advantages of farmland fragmentation in terms of diversification of crop planting can greatly increase farmers' agricultural income levels (Liu, 2017).

Secondly, farmland fragmentation promotes sufficient use of surplus rural labor. The increase in labor input promotes an increase in its marginal income, so as to raise farmers' income. The agricultural industry is still a labor-intensive industry in China. The agricultural production process requires a large amount of labor input to maintain production. There are a large number of surplus labors in rural areas in China. Therefore, in the context of farmland fragmentation, rural surplus labor can be effectively released. With the increase of labor input in the agricultural production process, the marginal income of labor gradually rises, and thus the agricultural income of the whole household will also improve step by step (Liu, 2011).

Thirdly, farmland fragmentation can promote the rational use of capital, thereby increasing farmers' agricultural income. Since most households with fragmented farmland often adopt diversified planting methods, for some plots, therefore, capital investment has reached a reasonable level, that is, as capital continues to be invested in the plot and when it reaches certain level, its unit income will decrease accordingly. Therefore, the limited capital of peasant households will be invested in plots that have not yet reached the stage of diminishing marginal output. The marginal income of the plots will increase significantly, thus increasing farmers' agricultural income (Zhang, 2008).

In addition, farmland fragmentation increases the probability of farmers engaging in non-agricultural production so as to make higher non-agricultural income. Over a constant total area of arable land of a peasant household, the severer the farmland fragmentation, the more labor the peasant household will have to input in the production process, in order to reach the same income level. Since the rate of return on labor in agricultural production is far less than that in non-agricultural industries, the severer the farmland fragmentation, the more surplus labor in peasant households will be transferred to non-agricultural industries, thereby increasing their non-agricultural income. Due to the advantage of higher rates of return on labor in non-agricultural industries, the peasant households will have higher income accordingly (Xu, 2011).

(3) Impact based on the perspective of land transaction and use

1) Farmland fragmentation and land transaction

The impact relationship between farmland fragmentation and land transaction is more reflected in the relationship between the scale of a single plot and the price of land. Many scholars held that the smaller the plot, the lower the transaction price of farmland (Kjelland, 2007; Latruffe, 2014). On the one hand, smaller plots are less likely to be used for agricultural investment or innovative production mode. Therefore, larger plots are usually more popular in the land transaction market. Sklenicka (2014) analyzed the transaction price of farmland in the land transaction market of Czech in 2008, found out that the transaction price of 1 *ha* of farmland on average is 1.44 times that of 0.4 *ha*, and the transaction price of 2 *ha* and 8 *ha* are 2.2 times and 2.8 times that of 1 *ha* respectively. On the other hand, due to the relatively poor infrastructure, including roads, for smaller plots, the outdated supporting infrastructure limits the transaction price of farmland. For example, Sklenicka (2014) stated that the transaction price of farmland plots accessible by roads in the Czech Republic is 2.1 times that of inaccessible plots. Some scholars also argued that fragmented plots also had certain impacts on the transaction price of surrounding critical plots. If the holders of fragmented land expand their production and operation scale through farmland transfer or lease, they have to transfer or lease the surrounding critical plots. Therefore, limited options may result in relatively high transaction prices for surrounding critical plots (Kjelland, 2007). Extant researches also presented that fragmented land plots were frequently traded in the land market, that is, their ownership was of poor stability (Sklenicka, 2014; Krusekopf, 2002). Normally, farmland holders only keep better plots for farming, while the plots with relatively poor quality and conditions are often leased or transferred to other farmers. Due to poor soil fertility conditions and weak production capacity, these plots may be frequently transferred.

2) Farmland fragmentation and sustainable land use

Farmland fertility is an important manifestation of sustainable land use, which determines the productivity of the farmland. Extant researches have shown that farmland fragmentation affects the fertility of farmland to a certain extent. Niroula (2005) and Jacoby (2002) believed that since farmers only concentrate on larger and closer plots for agricultural production, this will gradually reduce the soil fertility of the plots; in addition, farmers holding fragmented land usually do not apply fertilizers into all the plots, also causing the soil fertility of some plots to decrease. In addition to the impact from farmers' own production and operation on the fertility of the fragmented land, the production and management behavior of the lessee will also have a certain impact. Sklenicka (2014) discussed that because plots with poor fertility are most often leased to others, the tenants' insufficient investment and attention to water resources protection would result in less sustainable production capacity of the fragmented land. Sklenicka also confirmed the above viewpoint through research, and mentioned that the frequency of soil improvement by Czech landholders was 1.9 times that of farmland tenants.

In addition to the fertility of farmland, the impact of farmland fragmentation on land use is also a research focus. Many believed that fragmented land was easier to be converted for non-agricultural purposes (Deininger, 2012). In the process of agricultural production and operation, due to objective constraints, such as the scale, number of plots, and distance of the fragmented land, farmers tend to choose some of the plots for farming, leaving another some abandoned. Then these plots will gradually be converted for non-agricultural purposes (Dirimanova, 2010; Abubakari, 2016). For example, Qiu (2015) carried out researches on the Edmonton-Calgary region in Canada and confirmed the above view. Some also pointed out that the transfer of urban population into rural areas was the main reason for the conversion of fragmented land for non-agricultural purposes. For example, Sikor (2009) discussed that as the population of Albania continued to move into rural areas, resources such as capital and technology also flowed to the countryside, and the abandoned and fragmented farmland first became investment choices for non-agricultural purposes.

(4) Impact based on the perspective of eco-environment

As an important natural resource, land is closely related to the eco-environment. Some believed that farmland fragmentation could help enhance the overall eco-environment, more by protecting and maintaining biodiversity (Hartvigsen, 2014). In the process of agricultural production, in order to achieve the optimal yield, farmers make the best choice of crop types according to the fertility conditions of each plot. By planting different types of crops in different plots, the biodiversity is increased, which plays an important role in maintenance and improvement of the overall eco-environment (Latruffe, 2014). Based on researches on the relationship between farmland fragmentation and the eco-environment in Tijuana basin, Farley (2012) concluded that farmland fragmentation in the basin could keep accelerating the diversified development of biology and vegetation types in the region. Besides, some further stated that the existence of biodiversity had improved the ability of farmers to resist natural disaster risks, such as diseases and insect pests, and is conducive to further enhancing their production income (Di, 2010).

For the relationship between farmland fragmentation and the eco-environment, some also put forward different views, holding that farmland fragmentation had a negative impact on the eco-environment (Kjelland, 2007). For example, the more ownership holders within a certain area, the more difficult it will be to formulate ecological protection policies, which is not conducive to the protection of biodiversity to a certain extent. The fact that farmers with fragmented arable land usually choose to cultivate on a few fixed plots accelerates the depletion of water resources in the area. Farmland fragmentation leads to a decrease in native plant species and an increase in exotic species, which in turn accelerates soil degradation. It also leads to an increase in agricultural construction facilities, to damage the rural natural landscape.

2.1.1.5 Solutions to the Farmland Fragmentation

(1) Solutions to the farmland fragmentation all over the world

Although farmland fragmentation has a certain positive effect on agricultural production, land use and eco-environment, in general, most literatures show that the negative effect it brings is far from enough to match the positive effect. Thus, how to solve the problem of farmland fragmentation has become a hot-debated issue. Sikor (2009) held that land ownership adjustment was the most effective measure to solve the problem of farmland fragmentation. For example, in the early 1990s, despite a round of land reform in Central and Western Europe, farmland fragmentation over 13 countries in the region was still very serious. The area of a single farm was about 1 to 2.5 *ha*, and was divided into several extremely irregularly shaped plots. The fragmented land reduced the competitiveness of agricultural producers in the market and greatly restricted the agricultural development in the region. Therefore, shortly after the land reform, the Central and Western Europe region carried out another adjustment of land ownership (Riddell, 2002), which achieved significant results with the support of the Food and Agriculture Organization of the United Nations. The problem of farmland fragmentation has been resolved to a great extent (Sikor, 2009).

Some also believed that land consolidation led by agricultural engineering was another important measure to solve the problem of farmland fragmentation. It is one of the important means to solve the problem of farmland fragmentation, especially the natural farmland fragmentation, to conduct comprehensive improvement of farmland, upgrade basic supporting infrastructure, improve farmland quality, and thus improve agricultural production efficiency. For example, Janus (2017) have studied the effects of land consolidation in Poland and concluded that land consolidation could expand the area of effectiveness for fragmented land, thus exerting the effect of scale. Thereby, some scholars further studied the preference of farmers in solving the problem of farmland fragmentation through land consolidation. For example, Abdollahzadeh (2012) studied the preference of farmers in central Iran for land consolidation methods through field surveys, and found out that farmers more tend to participate in cooperative organizations for land consolidation. In addition, Latruffe (2014) held that since the area, number of plots, shape and other dimensions in the evaluation of farmland fragmentation had different effects on agricultural production, when carrying out land consolidation projects, it was necessary to focus on distinguishing the different dimensions of farmland fragmentation, and choose the most important dimension to solve the problem of farmland fragmentation.

In addition to the above institutional and engineering measures, some scholars also sorted out the specific practices in some countries. For example, Kawasaki (2010) explained some of the practices for moderating farmland fragmentation in Japan: 1) Providing subsidies for transfer of adjacent plots. If farmers need to increase the area of farmland, and the land to be transferred is geographically close to the land they own, the government will provide certain subsidies. The larger the transferred area, the greater the subsidies will be. 2) Providing subsidies for land consolidation. The government provides a cost subsidy of up to 30% to farmers who carry out land consolidation. 3) Supporting intermediaries. The government provides wages and training to intermediaries who facilitate land transactions, in order to give play to the role of intermediaries in the transference of neighboring plots. 4) Supporting joint production. The government promotes joint production of small-scale farmers with connected plots under different ownerships.

(2) Solutions to the farmland fragmentation in China

In the current process of farmland fragmentation governance, there are such problems as fragmentation of governance subjects, fragmentation of governance mechanisms, fragmentation of governance perceptions, fragmentation of governance social security, fragmentation of governance policies and laws. The main approach to solve the various difficulties in the process of farmland fragmentation governance is to carry out holistic governance, strengthen the overall understanding of farmland fragmentation, and build a coordinated governance mechanism between the government and the society (Wang, 2016). Current farmland fragmentation governance is mainly carried out by two means: land transfer and land consolidation.

1) Farmland ownership transfer

Some researches presented that farmland ownership transfer helped to reduce farmland fragmentation. Zhang Qinlan (2017) studied the impact of land transfer on the improvement of farmland fragmentation, and proposed that land transfer had a significant positive effect on lowering the degree of farmland fragmentation. Some scholars also pointed out that the effect of land transfer on reducing farmland fragmentation was not obvious. Xia Zhuzhi (2014) pointed out that it was difficult for the current spontaneous land transfer to overcome farmland fragmentation, and administrative intervention had even damaged the rights of disadvantaged farmers to trade independently. Therefore, the agricultural land transfer system should be innovated, in order to solve the problem of farmland fragmentation. Zhong Funing (2010) argued that due to the current mismatch of plots and excessively long transaction chains, it is difficult for farmers to transfer farmland successfully. At the current stage, only a substantial reduction in the number of rural households and permanent migration to cities can increase the number of plots and land areas in the hands of farmers, thereby increasing the probability of land matching and neighboring, expanding land segmentation, improving land substitution, and shortening transaction chains, reducing the transaction cost of land transfer between farmers, and achieving the goal of reducing the degree of farmland fragmentation through land transfer.

2) Land consolidation

Some scholars stated that land consolidation helped to improve farmland fragmentation. Deng Yao (2017) studied the impact of the field construction project on farmland fragmentation governance, and found out that the field construction project could generally improve the farmland fragmentation by 7%–24%, and the size and accessibility of the field are key indicators affecting the farmland fragmentation governance. Some also put forward different research findings. For example, Wen Gaohui (2016) used the synthetic index method and the binary logistic regression model to study the effect of land consolidation on farmland fragmentation in three districts and counties of Hubei Province. The study showed that the implementation of land consolidation projects could help lower farmland fragmentation, but with less obvious effect. In the implementation process of the land consolidation project, only by the combination of land consolidation and land ownership adjustment can farmland fragmentation be effectively controlled. Dai Shaoqi (2015), based on administrative villages and patches as evaluation units, studied the priority of land consolidation in the hilly and mountainous areas of Songxi County, Fujian Province, and found out that areas with low degree of farmland fragmentation and better location conditions should first go through land consolidation.

2.1.1.6 Review

In general, the extant literature has systematically summarized the concept and causes of the problem of farmland fragmentation. The evaluation methods of farmland fragmentation have been developed to be more mature and stable. The research on the impact of the problem of farmland fragmentation also shifted from a single research to a systematic research. The overall research on farmland fragmentation has been increasingly improved. However, the research on the problem of farmland fragmentation in China features a certain historical background. As a country long based on smallholders as agricultural production units, China is faced with severer farmland fragmentation and many challenges such as food security and sustainable agricultural development. Therefore, we should fully draw from the research approaches and perspectives of foreign scholars on the problem of farmland fragmentation, and continue to deepen and explore relevant researches on the problem of farmland fragmentation in combination with the actual situation of China.

First, we should deepen the research on the evaluation methods for farmland fragmentation. China boasts a vast territory, with huge differences in natural conditions, historical backgrounds, and social culture in various regions. The causes of farmland fragmentation are different from each other, and therefore the degree of fragmentation also turned out to be different. The sole application of economic or ecological methods in evaluating the degree of farmland fragmentation fails to fully reflect the actual situation of farmland fragmentation within a region, which may lead to deviations in the results of impact research or the response to the problem of farmland fragmentation. Therefore, we should fully draw from foreign researches to continuously innovate the evaluation methods of farmland fragmentation, and then establish an evaluation method suitable for specific regional analysis.

Second, we should expand researches on the impact of farmland fragmentation. The researches in China related to the impact are mostly carried out in the field of agricultural production, such as the impact of farmland fragmentation on agricultural production costs, yield, production efficiency, and the use of new technologies. In comparison, in addition to the agricultural field, foreign researches have more transited to the evaluation of more comprehensive impacts, such as eco-environmental impacts, social impacts, and population migration impacts. Therefore, the transition from a single-field impact effect research to a multi-field systematic research should be an important topic for the Chinese researches in the future.

Third, we should improve the research on the "degree" of farmland fragmentation. In the context of the implementation of the "three rights separation" of rural land in China, it is also an important issue of concern in future researches to explore the level of reduction in the degree of farmland fragmentation in order to meet the sustainable agricultural development in various regions, and to achieve the profitable economy of scale.

Generally speaking, foreign academic communities have carried out more researches on the impact of farmland fragmentation and established specific research viewpoints, which will undoubtedly provide a good academic reference for the research on farmland fragmentation in China. First, we should focus on the systematic research on the impact of farmland fragmentation. In addition to analyzing economic benefits, the research on the impact of farmland fragmentation also needs to focus on environmental benefits, analyze the impact of farmland fragmentation in the economic environment system, and then put forward reasonable suggestions to solve the problem of farmland fragmentation. This should be a new trend for the future researches. Second, we should pay attention to the research on the "degree of farmland fragmentation". Foreign scholars found out that due to the different causes of farmland fragmentation, excessively high or low fragmentation in the agricultural production process would have an adverse impact. However,

CHAPTER 2
LITERATURE REVIEW AND THEORETICAL BASIS

China has a vast territory with huge differences in natural, economic, social and cultural fields. Therefore, the degree and impact of farmland fragmentation are bound to be various in different regions. The future research should focus on the degree of farmland fragmentation in different regions.

2.1.2 Rural Irrigation Collective Action

2.1.2.1 Evaluating the Capacity of Rural Irrigation Collective Action

(1) Perspectives for evaluating the capacity of rural irrigation collective action

At present, many evaluation researches have been carried out on the issue of rural irrigation collective action, to present quite fruitful findings. Generally speaking, current scholars have different perspectives on the evaluation and research of rural irrigation collective action. For example, the supply of small-scale farmland water conservancy facilities in rural areas is used as an index for evaluating rural irrigation collective action, that is, to adopt the supply of public goods as a standard for measuring rural irrigation collective action (Cai and Zhu, 2016; Cai and Cai, 2014; Miao, 2014). In addition, some scholars also evaluated rural irrigation collective action from their own points of view, which can be generally divided into two perspectives. The first perspective is the actual behavior of farmers as rural irrigation collective action. For example, the performance of the current rural participation in management and protection of farmland irrigation facilities is used as a standard for evaluating the capacity of rural irrigation collective action (Yang, 2018; Ren, 2016). The second perspective is the future behavior of farmers participating in rural irrigation collective action. For example, the willingness in the following terms of small-scale farmland water conservancy facilities can be adopted as indicators for evaluating the rural irrigation collective action, including the willingness of cooperative supply (Wang, 2012), construction (Ke, 2015), investment (He, 2014), cost sharing (Wang, 2014), and irrigation management reform (Xu, 2015).

Rural irrigation facilities are typical common-pool resources, that is, there is a competitive relationship among farmers in the process of using irrigation facilities. The water resources consumed by a single peasant household cannot be used by other peasant households. But at the same time, there are also non-exclusive characteristics, that is, if a peasant household does not participate in the construction or maintenance of irrigation facilities, the possibility of using irrigation facilities still cannot be excluded. All in all, no matter from which perspective the rural irrigation collective action is evaluated, the above indicators are in line with the characteristics of common-pool resources, and can well reflect the situation of rural irrigation collective action.

(2) Status quo of the capacity of rural irrigation collective action

The analysis of the status quo of the capacity of rural irrigation collective action is an important basis for the analysis of the influencing factors and mechanism underlying the capacity of rural irrigation collective action. At present, many Chinese scholars have conducted researches on the status quo of rural irrigation collective action in China based on big data, and the researches showed that the current capacity of rural irrigation collective action in China is rather low. For example, Cai Jingjing (2015) used the survey data of 430 farmers in 15 provinces to analyze the willingness of farmers to cooperate in irrigation, and found out that the willingness of farmers to cooperate in irrigation was low; Cai Qihua (2014) adopted the data of 1,024 farmers of Inner Mongolia Autonomous Region, Ningxia Hui Autonomous Region and Shandong Province to analyze the farmers' willingness to engage in rural irrigation collective action (supply of small-scale farmland water conservancy facilities), and statistical data showed that the farmers' willingness to participate in the supply of small-scale farmland water conservancy facilities was only 65%, and only 68% of these farmers choose to participate in the supply through investment; Wang Xin (2014) used the survey data of 890 peasant households in Shaanxi Province to study the influencing factors of rural irrigation collective action (the willingness to share the cost of cooperative supply of small-scale water conservancy facilities), and found out that the willingness of peasant households to participate in cost sharing was poor, with only 36.7% of peasant households showing the intention of participation; Ren Guizhou (2016) adopted the survey data of 420 peasant households in southern Jiangsu to analyze the problems existing in rural irrigation collective action (management and protection of farmland water conservancy facilities), and concluded that the current participation of farmers in the management and protection of small-scale farmland water conservancy facilities suffered from such problems as low participation, weak coordination and low efficiency; Cai Rong (2015) used 129 samples of farmers in Nanyang Town, Yancheng City, Jiangsu Province to analyze the status of rural irrigation collective action (cooperative supply of small-scale farmland water conservancy facilities) and its influencing factors, and pointed out that farmers were less willing to engage in the construction and maintenance of small-scale farmland water conservancy facilities, with less than 40% of the farmers showing the willingness.

2.1.2.2 Influencing Factors of Rural Irrigation Collective Action

(1) Qualitative analysis

Since the capacity of rural irrigation collective action is at a relatively low level in China, the analysis and research on the underlying reasons become very important. Generally speaking, some scholars analyzed the reasons for the difficulty in forming rural irrigation collective action in China from qualitative perspectives. For example, Ding Jianjun (2012) studied the problems in the irrigation collective action in Wangping Village, Shayang County, Hubei Province, and found out that the reasons for the difficulty in forming rural irrigation collective action included the disintegration of village-level organization, the "atomization" among villagers and the difficulty in implementing the system of "specific negotiation for each case", and the weak village-level organizational capacity. Wu Qiuju (2017) adopted game theory to study farmland water conservancy cooperation among farmers, and concluded that the spontaneous cooperation among farmers in farmland water conservancy may not be achieved, mainly due to the weakening of the rural community capacity composed of integration, participation and execution capacities, which fails to effectively resolve the participation costs, integration costs and execution costs; specifically, the reason for the weakening of community capacity is mainly the atomization of rural society and the virtualization of rural grassroots organizations.

(2) Quantitative analysis

In addition to the above qualitative analysis, most scholars mainly adopt quantitative methods to analyze the influencing factors of rural irrigation collective action. In general, current researches mainly use field surveys to obtain relevant data, and on this basis adopt regression analysis, factor analysis and other methods to analyze the extent to which each element affects rural irrigation collective action. From the extant literatures, some scholars analyzed all the relevant elements that affect the rural irrigation collective action, and proposed the most important influencing factors. For example, Cai Jingjing (2015) pointed out that the education level of farmers and the initiative to resolve water disputes were the main factors that promoted farmers to cooperate for irrigation. Cai Qihua et al. (2014) argued that there was a certain positive correlation between the social trust and income level of farmers and the supply of rural irrigation facilities. Cai Rong et al. (2014) held that the impact from village size and income inequality on rural irrigation collective action both presented "inverted U-shaped" characteristics. They also pointed out that there was an inverted U-shaped relationship between endowment heterogeneity and collective action, and moderate endowment heterogeneity contributed to the success of collective action; social heterogeneity had a negative effect on collective action; and the gender of community members also had a certain impact on collective action. Researches by Ke Xinli (2015) revealed that age, education level, proportion of agricultural income, evaluation of facility water yield, and current management and protection are the main factors affecting rural irrigation collective action. Wang Xin (2014) believed that farmers' cognition degree in terms of water conservancy had the greatest positive effect on rural irrigation collective action, followed by the expected rate of return, while risk preference and basic characteristics of peasant households had little effect.

In addition to the above researches, some scholars also analyzed specific influencing factors. For example, Ding Dong (2013) analyzed the impact of economic heterogeneity, social and cultural heterogeneity, intergenerational gender heterogeneity, and political heterogeneity on rural irrigation collective action, using survey data from 400 peasant households in Hubei Province. It was indicated that social and cultural heterogeneity and political heterogeneity were important factors that affect rural irrigation collective action, while economic heterogeneity and intergenerational gender heterogeneity had little effect. Cai Qihua (2017) used the data of 1,440 peasant households in Ningxia Hui Autonomous Region, Shaanxi province, and Henan province to analyze the impact of the relationship network on farmers' participation in rural collective action (investment in the construction of small-scale farmland water conservancy). In terms of relationship network evaluation, the "number of relatives and friends owned by the household" is used as an index for evaluating the relationship network, the "number of friends owned by the household" is used as a weak index for evaluating the relationship network, and the "number of relatives owned by the household" is used as a strong index for evaluating the relationship network. The research showed that the relationship network could promote the success of rural irrigation collective action. Miao Shanshan (2014) studied the impact of social capital on rural irrigation collective action (cooperation on small-scale water conservancy facilities), and argued that the trust dimension, participation dimension, and network dimension were the main factors driving farmers to engage in rural irrigation collective action, and the prestige dimension reduced farmers' participation; in addition, the multi-dimensional heterogeneity of social capital would also lead to uncertainty in the results of collective actions. Xu Lang (2015) studied the impact of social capital on rural irrigation collective action (willingness for irrigation management reform), and held that social trust and social network had positive effects on the development of rural irrigation collective action, but the impact from social norms was not obvious. Wang Xin (2012) studied the impact of social capital on rural irrigation collective action (total willingness for cooperation on small water conservancy facilities), and further argued that social capital played a positive role in promoting rural irrigation collective action. Wang Huina (2013) studied the relationship between group characteristics and self-organization and governance for irrigation, and found out that group homogeneity facilitates the formation of rural irrigation collective action.

2.1.2.3 Influencing Mechanism of Rural Irrigation Collective Action

There are many factors affecting the rural irrigation collective action, and different factors play different roles in the formation of rural irrigation collective action. The above research results show that some factors can directly promote or inhibit the success of rural irrigation collective action, while some have no influence on the formation of rural irrigation collective action. However, there are also complex influence relationships among different factors. Therefore, it is of great guiding significance to explore the formation mechanism of rural irrigation collective action and analyze how a certain factor influences the rural irrigation collective action through influencing other factors. After combing the existing research literature, it is found that the perspectives and emphases of the existing researches are different. For example, Guo Zhen (2015) studied the impact of village size and farmland ownership circulation on rural irrigation collective action. The study pointed out that village size and membership structure are important factors affecting rural irrigation collective action, but the farmland ownership circulation could effectively change village size and membership structure, gradually reducing operators on the one hand, and making the membership structure become heterogeneous on the other hand, thus promoting the success of rural irrigation collective action. Yang Liu (2018) studied the impact of social trust and organizational support on rural irrigation collective action (the performance of water resources management). This study believed that social trust and organizational support could significantly positively affect the performance of farmers' participation in management and care for small-scale farm hydro project. Organizational support can enhance the positive effect of social trust on the performance of water resources management. Organizational support exerts an impact on the performance of water resources management through influencing farmers' recognition of village cadres and the management and care system. Cai Qihua (2016) pointed out in their studies that social capital can positively influence the willingness and degree of farmers to engage in rural irrigation collective action, and income gap has a positive impact on the willingness and a negative impact on the degree. In terms of cross-influence, social capital can reduce the negative impact of income gap on rural irrigation collective action.

2.1.2.4 Regional Comparative Analysis of Rural Irrigation Collective Action

In addition to the analysis of the factors and mechanisms of rural irrigation collective action, some scholars also comparatively analyzed the regional conditions of rural irrigation collective action. For example, Cai Jingjing (2015) believed that the cooperation willingness of farmers in inland areas is stronger than that in coastal areas, and the cooperation willingness of farmers in developed areas is stronger than that in backward areas. He Pingjun (2016) pointed out in their studies that farmers in major grain-producing areas were more willing to invest in farmland irrigation facilities than those in non-major grain-producing areas.

2.1.2.5 Review

Through combing the existing research literature, it is found that there are rich research results on rural irrigation collective action. For example, the evaluation index of rural irrigation collective action involves a wide range of fields, the external influencing factors of rural irrigation collective action have achieved relatively consistent research results, and the complex relationship among the influence of many key factors on rural collective action has been analyzed in the research of rural irrigation collective action mechanism. But at the same time, the above researches also have some shortcomings. First, the research form is simple. For example, many researches have adopted field investigation to collect research data of peasant household samples, and then established an index system for evaluating external factors affecting rural irrigation collective action. In addition, regression analysis and other methods have been used to analyze the direction and degree of influence of each factor, and then the most important factors have been pointed out. This kind of research accounts for the vast majority of the current researches on the influencing factors of rural irrigation collective action. Second, there are few literatures on the analysis of the research mechanism. Different factors play different roles in the formation process of rural irrigation collective action, and there is a complex influence relationship between these factors, but there is little research on the mechanism in the existing literature. Third, there are few research literatures about the characteristic issue. Most of the literatures analyzed the impact of the age, education level, family income and other factors of the head of a peasant household on rural irrigation collective action, while there are few studies on the characteristics of China's regional characteristics such as labor outflow and land fragmentation.

2.2 Theoretical Basis

2.2.1 Institutional Analysis and Development Framework

Institutional Analysis and Development (IAD) framework is one of the competitive theories in the current theories of policy process which study public policy from the perspective of institution, and it is also the institutional theory which has the greatest influence on the policy process research (Li, 2016). Elinor Ostrom, the founder of the Institutional Analysis and Development framework, was also awarded the 2009 Nobel Prize in Economics for her outstanding contributions to the governance of public resources. Since 1982, the Institutional Analysis and Development framework has been one of Elinor Ostrom's major studies. The framework is mainly used to explain how external variables, including application rules, affect policy outcomes in autonomous governance of common-pool resources, so as to provide a set of institutional design schemes and standards that can enhance trust and cooperation among resource users (Poteete, 2010).

On the whole, a complete set of Institutional Analysis and Development framework mainly includes seven components (see Figure 2–1).

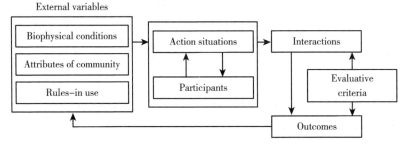

Figure 2 –1 Institutional Analysis and Development framework

In general, the key to the application of Institutional Analysis and Development framework is to identify how the action situations and participants interact with each other in the action arena under the influence of external variables, and what kind of results this interaction may produce, as well as how the result may react on both. The Institutional Analysis and Development framework indicates that the factors affecting the action arena mainly include three aspects: biophysical conditions, attributes of community and rules-in use.

The biophysical conditions are the main factors that affect the action arena. Biophysical conditions will act on a series of factors such as actions and results in the action arena. Participants' actions under different institutional rules are all based on the existing biophysical conditions. For a long time, people have usually adopted Paul A. Samuelson's dichotomy to classify goods, that is, goods mainly include private goods and public goods. In 1965, James M. Buchanan further put forward the third type of goods, namely club goods. In 1977, Vincent Ostrom further classified the above three goods and proposed a fourth type, the common-pool resources. According to the competitive and exclusive nature of the goods, the four types of goods are classified as shown in Table 2–1.

Table 2–1　　　　　　　　　Classification of goods

		Subtractability of use	
		Low	High
Difficulty of excluding potential beneficiaries	Low	Toll goods	Private goods
	High	Public goods	Common-pool resources

Source: *Elinor Ostrom, Understanding Institutional Diversity, Princeton University Press,* 2005.

The attribute of community is another important factor that affects the action arena. In general, the differences among different members in the community will affect the decision-making process. For example, the heterogeneity of different members in rural communities in terms of economy, education, social capital and other aspects will often affect the development of rural collective action. Therefore, the unequal distribution of the economic and social resources of the community members often affects the interaction and behavior results in the action arena.

Rules-in-use is one of the important variables that influence the action arena, but the study of institution is faced with great challenges, such as the diversity, unpredictability, multidisciplinary and multilevel of the institution. Elinor Ostrom thought institution influences action mainly through the structural framework of the action scenario. Therefore, she proposed seven groups of rules for rule structure, as shown in Figure 2–2.

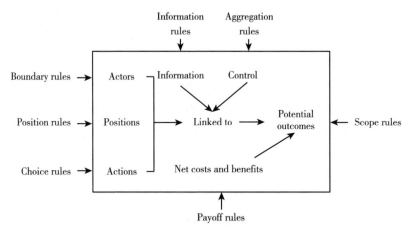

Figure 2 –2 Rules as exogenous variables directly affecting the elements of an action situation

Source: *Elinor Ostrom, Background on the Institutional Analysis and Development Framework, Policy Studies Journal*, 2011(39).

2.2.2 Social-ecological System Framework

In 2007, Elinor Ostrom developed the Social-ecological System framework, which attracted a lot of attention from the international social science community (see Figure 2 – 3). The Social-ecological System framework is mainly composed of resource system, resource units, governance system, users, social, economic, and political settings and related ecosystems (Ostrom, 2007). Among them, the resource system, the resource units, the governance system and the users are the four core subsystems, while the social, economic and political settings, and the related ecosystems are the external variables. Each of these subsystems forms the first layer of the Social-ecological System framework. In addition, in each subsystem, there are complex sub-variables, which are the second layer of the framework (see Table 2 – 2). Scholars can further develop the third-level or fourth-level variables according to their own research needs (Ostrom, 2007).

With multiple levels, the Social-ecological System framework is possible to make an adequate diagnosis of the entire social ecosystem. For example, scholars can fully combine the Social-ecological System framework with specific cases, and make full use of variables in the Social-ecological System framework to connect, so that the Social-ecological System framework can be effectively applied to the analysis of specific problems.

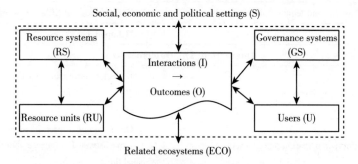

Figure 2-3　Social-ecological system(First layers)

Source: *Elinor Ostrom, A General Framework for Analyzing Sustainability of Social-ecological Systems, Science,* 2009(7).

Table 2-2　　　Social-ecological system(Second layers)

Social, economic and political settings(S)
S1-Economic development; S2-Demographic trends; S3-Political stability; S4-Government resource policies; S5-Market incentives; S6-Media organization; S7-Technology

Resource Systems(RS)	Governance Systems(GS)
RS1 Sector (e.g., water, forests, pasture, fish)	GS1 Government organizations
RS2 Clarity of system boundaries	GS2 Nongovernment organizations
RS3 Size of resource system	GS3 Network structure
RS4 Human-constructed facilities	GS4 Property-rights systems
RS5 Productivity of system	GS5 Operational rules
RS6 Equilibrium properties	GS6 Collective-choice rules
RS7 Predictability of system dynamics	GS7 Constitutional rules
RS8 Storage characteristics	GS8 Monitoring and sanctioning processes
RS9 Location	

continued

Resource Units (RU)	**Users (U)**
RU1 Resource unit mobility	U1 Number of users
RU2 Growth or replacement rate	U2 Socioeconomic attributes of users
RU3 Interaction among resource units	U3 History of use
RU4 Economic value	U4 Location
RU5 Number of units	U5 Leadership/entrepreneurship
RU6 Distinctive markings	U6 Norms/social capital
RU7 Spatial and temporal distribution	U7 Knowledge of SES/mental models
	U8 Importance of resource
Interaction (I) → Outcomes (O)	
I1 Harvesting levels of diverse users	O1 Social performance measures (e.g., efficiency, equity, accountability, sustainability)
I2 Information sharing among users	O2 Ecological performance measures (e.g., overharvested, resilience, bio-diversity, sustainability)
I3 Deliberation processes	O3 Externalities to other SESs
I4 Conflicts among users	
I5 Investment activities	
I6 Lobbying activities	
I7 Self-organizing activities	
I8 Networking activities	

Related Ecosystems (ECO)
ECO1-Climate patterns; ECO2-Pollution patterns; ECO3-Flows into and out of focal SES

Source: *Elinor Ostrom, A General Framework for Analyzing Sustainability of Social-ecological Systems, Science,* 2009 (7).

CHAPTER 3

LOGICAL RELATIONSHIP BETWEEN FARMLAND RESOURCES AND RURAL IRRIGATION COLLECTIVE ACTION

As a typical natural resource, farmland resource plays an important role in the process of agricultural production. Therefore, in the formation of rural collective action, farmland resource is bound to have a certain influence. Small-scale peasant households are the main participants of rural collective action. Different characteristics of small-scale peasant households have different effects on rural collective action from many perspectives. In this chapter, starting from the perspective of small-scale peasant households and on the basis of combing the historical evolution of small-scale peasant households' participation in rural collective action, the logic of small-scale peasant households' participation in rural collective action is analyzed from the perspective of the characteristics of small-scale peasant households. Secondly, due to the lack of resource endowment in the characteristics of small-scale peasant households, this book analyzes the impact of farmland resources on rural collective action from different perspectives. Finally, the logical relationship between farmland fragmentation and rural irrigation collective action is analyzed.

3.1 Collective Action of Small-scale Peasant Households

3.1.1 Historical Evolution of Participation of Small-scale Peasant Households in Rural Collective Action

Since 1949, the development of small-scale peasant households in China has experienced three stages. The first stage lasted from 1949 to 1953. In 1950, the Eighth Session of the Committee of the Central People's Government passed the *Land Reform Law of the People's Republic of China*, which proposed to abolish the land ownership of feudal landlords and implement the land ownership for small-scale peasants. By the end of 1952, 480 million farmers in the new and old liberated areas had acquired ownership of more than 700 million *mu* of land. The second stage lasted from 1953 to 1978. At this stage, land and other means of production were collectively owned by peasants, and the structure of small-scale peasant economy was completely broken after the exploration from "large size and high-degree of socialization of commune" to "the means of production and products owned by the commune, the production brigade and the production

CHAPTER 3
LOGICAL RELATIONSHIP BETWEEN FARMLAND RESOURCES AND RURAL IRRIGATION COLLECTIVE ACTION

team respectively, and the production team level serving as the basic accounting unit" in the period of people's commune. The third stage is the reform and opening up since 1978. The current situation of small-scale peasant households in China is mainly attributed to the changes in China's rural management system around the 1980s. With the establishment and implementation of the Household Contract Responsibility System, more than 200 million peasant households nationwide have become the most basic production and management units, and more than 4.5 million original rural management and accounting units have been canceled (Su, 2017). At this stage, the reform of the land system enabled farmers to have the right to contract and manage farmland, which stimulated the production vitality of small-scale peasant households and made them the most basic production and management units in China.

The small-scale peasant is the production and living unit with the family as the carrier. It is not only the basic unit of agricultural production in China, but also the basic member of the rural community (Zhang, 2019). The participation, recognition and support of small-scale peasant are the intrinsic power to maintain community stability and promote community development. Therefore, with the changes of agricultural management subjects in different stages, the governance subjects and capabilities of rural public affairs in China present different characteristics. Before the land reform, the patriarchal system was the main system that facilitated collective action in rural areas, in which the clan elders had absolute control. With the land reform from 1949 to 1953, this traditional rural social governance structure changed. Under the leadership of the Communist Party of China, new social relations had been established in the rural areas, and the rural leaders had been reshaped, as well as the characteristics of administrative management of rural governance had been embodied, and the corresponding institutional system has promoted the formation of rural collective action. However, since 1978, with the establishment of the Household Contract Responsibility System, political and social affairs began to be separated, and rural public affairs governance was gradually implemented by village collective organizations, weakening the capacity of rural collective action.

The construction of farmland irrigation facilities in China has experienced a tortuous development process (Wang, 2013). In the 1950s, under the people's commune system with the commune serving as both an administrative organization and an economic organization, collective organizations organized peasant households to build irrigation facilities, which was also the basis of the construction of water conservancy facilities in China. After 1978, the government power gradually withdrew from the village, and the village could continue to organize farmers' families to build farmland irrigation facilities to some extent. However, due to the withdrawal of government power, the organizational capacity became weak. After the reform of rural taxes and fees in 2003, the "compulsory labor" and "accumulated labor" systems were abolished completely, making it difficult for village collectives to organize villagers to build and maintain irrigation facilities. Although the government would invest in the construction of irrigation facilities, this investment was generally limited to the large and medium-sized irrigation facilities, while the construction and maintenance of small irrigation facilities were mainly carried out by owners and users. In general, the ownership of small irrigation facilities belonged to the village collective, and the right to use belonged to the villagers. Since it is difficult for the village collective to organize the villagers to carry out the construction and maintenance of irrigation facilities, therefore, from 2003 to now for more than ten years, the original construction of farmland irrigation canal basically fell in the state of no money to repair and no one to take care of, with serious damage

(Wu, 2017). At the same time, some problems arose, such as high irrigation costs for peasants and lower crop yields.

From the development of small-scale peasant households' participation in irrigation facilities, the basic unit of rural irrigation collective action in China has been narrowed from the original people's commune to the village, and further to a few peasants or a single peasant in recent years; the management of public good has shifted from collective cooperation to individualization, and the capacity of rural irrigation collective action in China has been comprehensively reduced. Seemingly, the process of small-scale peasant households participating in the collective action for rural irrigation is actually a process in which a large number of household-based small-scale peasant households make up the community and choose to engage in the collective action independently under the guidance of the government. Furthermore, its essence is that the competitive and non-exclusive nature of rural common-pool resources leads to the inconsistency between the individual choice and the collective choice of small-scale peasant households, and the "free-riding" phenomenon of small-scale peasant households' participation in the supply of public goods is widespread, which ultimately leads to the dilemma of collective action (Xu, 2015; Cai, 2014).

3.1.2 Logical Relationship between the Characteristics of Small-scale Peasant Households and Rural Collective Action

3.1.2.1 Resource Endowment and Rural Collective Action

A big country with a large population of smallholders is the fundamental realities in China, and the shortage of per capita resource endowment is the most basic feature of China's small-scale peasant households. According to

FARMLAND FRAGMENTATION AND COLLECTIVE ACTION:
A STUDY ON THE IRRIGATION SYSTEM IN CHINA

the data released by the World Bank, the world's per capita farmland area was 0.196 ha in 2014, while China's per capita farmland area was only 0.077 ha. In addition, in terms of water resources, China's per capita water resource was 2,173 cubic meters, only one quarter of the world's per capita level. This shows that the character of insufficient per capita resource endowment has become the primary marker of small-scale peasant households in China. As the most basic means of agricultural production, soil and water resources have a relatively important effect on rural collective action. The lack of per capita resources makes it difficult to construct the agricultural industrial system, production system and management system, which affects the degree of small-scale peasant households' participation in rural collective action from many aspects, especially the supply of basic public goods in agricultural production.

The characteristics of insufficient per capita resource endowment of small-scale peasant households have positive and negative effects on rural collective action. First of all, this characteristic can have a negative impact on rural collective action. Small-scale peasant households' per capita resource endowment is insufficient, which means that there are more peasants in a certain area. According to the research of Olson (1965), the larger the scale, the more difficult it is for the members participating in collective action to form joint forces, thus the collective action capacity decreases. For example, with regard to the maintenance of farmland irrigation facilities, the larger the number of small-scale peasant households in the region, it is difficult to reach agreement on the water distribution system, irrigation maintenance system, payment system and other aspects and form a unified system and rules. Therefore, it is difficult to form collective action in the end. Secondly, this characteristic promotes the formation of rural collective action from another angle. With a small per capita resource endowment, small-scale peasant households may face high-cost pressure in the construction and maintenance of infrastructure. Therefore, in order to reduce the cost of agricultural production, small-scale peasant households may reduce the cost by participating in collective action. For example, for households with less farmland resources, the investment cost caused by the construction and maintenance of infrastructure such as roads and irrigation facilities is relatively high for a single small-scale peasant household. At this time, small-scale peasant households usually choose to reduce the total cost of agricultural production by participating in collective action.

3.1.2.2 Peasant Households Differentiation and Rural Collective Action

With the continuous development of urbanization in China, the rural labor force keeps flowing out, and the income structure of small-scale peasant households keeps changing, and the proportion of the total agricultural income in the total household income keeps declining, that is, it shows the evolutionary characteristics of full-time farmers → part-time farmers → non-farmers. According to the calculation of Zhang Chen, if the proportion of total agricultural income to total household income is more than 80% and that of total agricultural income is less than 20% as full-time farmers and non-farmers respectively, then from 2003 to 2016, the proportion of full-time farmers, part-time farmers and non-farmers in China has changed from 11.18%, 55.54% and 33.28% to 2.9%, 73.89% and 23.21%, respectively. With the continuous outflow of labor force, the income of peasant households has diversified characteristics, and the income difference among peasant households has gradually widened, and the development of rural collective action is bound to be affected to some extent (Zhang, 2019).

FARMLAND FRAGMENTATION AND COLLECTIVE ACTION:
A STUDY ON THE IRRIGATION SYSTEM IN CHINA

The effect of peasant household differentiation on rural collective action is also dualistic. On the one hand, the differentiation of peasants means that the income level of some small-scale peasant households has increased, and higher income level is good for the development of rural collective action. For example, in the construction and maintenance of rural roads or farmland irrigation facilities, a large number of funds need to be invested for maintenance. Therefore, a higher household income level is more conducive to promoting the participation of small-scale peasant households in rural collective action. On the other hand, the differentiation of peasants also indicates that the dependence of small-scale peasant households on agricultural production has decreased. In the process of the transformation of peasants from full-time farmers to part-time farmers and non-farmers, the proportion of agricultural income in the total household income decreases, and the participation degree of peasants in collective action related to agricultural production may decrease.

In addition to the above factors, there are many other characteristics of small-scale peasant households, which also affect the development of rural collective action in China. For example, due to the restriction of resource endowment conditions, it is difficult for small-scale peasant households to obtain a higher income. In China, the dual structure between urban and rural areas promotes the continuous transfer of small-scale peasant households from rural areas to urban areas, which leads to the decrease of small-scale peasant households' dependence on agricultural production, the weakening of rural social capital and the decline of rural leadership. The characteristics of small-scale peasant households themselves affect the rural collective action from all aspects. On the whole, the current characteristics of small-scale peasant households have a significant negative impact on China's rural collective action, that is, China's fundamental realities of a big country with a large population of smallholders lead to the formation of China's rural collective action.

3.1.3 Logic of Small-scale Peasant Households' Participation in Rural Collective Action based on IAD Framework

Small-scale peasant households are the main participants of rural collective action, and their multiple characteristics have an important impact on rural collective action. With the help of Institutional Analysis and Development (IAD) framework (Ostrom, 2005), the impact of the characteristics of small-scale peasant households on rural collective action can be shown in Figure 3-1.

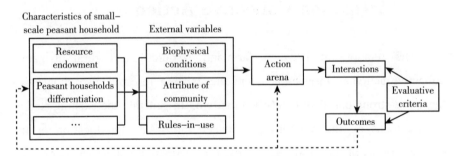

Figure 3-1 Logical relationship between the characteristics of small-scale peasant households and rural collective action based on IAD framework

According to the Institutional Analysis and Development framework, the external factors affecting rural collective action can be summarized into three aspects: biophysical conditions, attributes of community and rules-in-use. These three kinds of external variables constantly influence the action arena constituted by actors and action scenarios, and generate outcomes through interaction, which then response and influence external factors, and finally form a complete action system. Based on the above analysis of the characteristics of small-scale peasant households, it can be seen that the characteristics of small-scale peasant households, such as insufficient resource endowment and severe differentiation of farmers, can be summed up as one of the three types of external variables of collective action, and they are intersecting. In general, characteristics such as insufficient per capita resources and differentiation of peasant households can be classified into different conceptual categories in the external factors of Institutional Analysis and Development framework, and they themselves also affect rural collective action through a variety of intermediary mechanisms.

3.2 Farmland Resources and Rural Irrigation Collective Action

The relationship between farmland resources and rural irrigation collective action is inseparable, and the existing research also shows that there is a strong correlation between the two. In general, the existing studies mainly focus on the relationship between the scale, the location or the property rights of the farmland and the rural irrigation collective action.

3.2.1 Farmland Size and Rural Irrigation Collective Action

Most scholars believe that the larger the scale of farmland, the stronger the degree of farmers' participation in irrigation collective action (Araral, 2009; Sharaunga, 2017; Fitz, 2018). Collective action theory holds that heterogeneous groups are more likely to contribute to the success of collective action, because members of heterogeneous groups have stronger incentives to engage in collective action, and plot size is the key dimension of member heterogeneity.

Specifically, there are mainly the following reasons: first, the more farmland the family manages, the greater the water demand for farmland, and the stronger the peasant's dependence on water conservancy infrastructure. Therefore, peasants are more concerned about the construction and maintenance of agricultural irrigation facilities (Manjunatha, 2013). Second, under the condition of the same disaster intensity, the larger the farmland area of the household, the higher the economic losses suffered. Thus, in order to ensure irrigation conditions and improve agricultural production conditions, peasants with larger farmland area are more willing to engage in irrigation collective action (Cai, 2013). Third, peasants with a large scale of farmland have a relatively high level of household income (Dylan, 2018), so they are more able to afford the cost of construction and maintenance of irrigation facilities (Cai, 2014).

Some scholars also believe that the scale of farmland has a negative impact on the rural irrigation collective action, that is, the larger the area of farmland, the worse the degree of peasants' participation in the irrigation collective action (Cai, 2016). There are two main reasons for this. First, with the increase of farmland area, scale benefit drives peasants to purchase or lease small water conservancy facilities to meet their irrigation needs, so as to increase land marginal income and achieve the goal of maximizing income (Miao, 2014). Second, the larger the scale of farmland, the greater the capital input required for the construction of irrigation facilities, and the greater the cost in fertilizers, pesticides and other inputs, so farmers cannot afford the construction of irrigation facilities (Juan, 2018). In addition to the above studies, there are also scholars who believe that the area of farmland has no significant effect on the rural irrigation collective action (Cai, 2016).

3.2.2 Farmland Location and Rural Irrigation Collective Action

In general, the farmland location is mainly presented by two aspects. The first is the topography. According to the existing research, the terrain of farmland has an important effect on irrigation collective action. Some scholars believe that if the farmland is located in plains and other areas, due to the relatively flat terrain, the input cost of the construction and maintenance of collective irrigation facilities is relatively low (Panagopoulos, 2014), so the peasants will engage in the collective action to a higher degree. Some scholars believe that if the farmland is located in mountainous or hilly areas, although the construction and maintenance cost of collective irrigation facilities is higher, the cost will be relatively lower in the case of joint cooperation compared with the construction or maintenance of irrigation facilities by peasants themselves (Yuan, 2016). Therefore peasants are more inclined to take collective action (Gao, 2016).

The second is the distance between the farmland and the river. According to existing studies, the influence of the distance between farmland and river channel on irrigation collective action presents an inverted U-shaped structure. When the farmland is located near to the river, due to abundant water resources, the willingness of peasants to take irrigation collective action will be lower, because abundant water resources will weaken peasants' participation in the construction of irrigation facilities to expand water supply (Yuko, 2010; Ricks, 2016). Therefore, with the increase of the distance between farmland and water resources, the willingness of peasants to engage in irrigation collective action will be enhanced. However, when the distance of farmland reaches a certain level, the degree of peasants' participation in irrigation collective action will become worse, because if the distance is too far, the degree of water resource scarcity will be too high and the investment cost will be too high (Agrawal, 2005; Araral, 2009).

3.2.3 Farmland Property Rights and Rural Irrigation Collective Action

Existing studies have pointed out that the adjustment of farmland property rights is not beneficial for peasants to engage in irrigation collective action (Cai, 2017), because the change of farmland property rights affects the stability, and the construction of collective irrigation facilities requires a large amount of capital investment in the early stage. If the property right of farmland is not stable, the return rate of investment is low, so the unstable property right is not conducive to the collective action (Michael, 2014; Hanemann, 2014; Hausner, 2015). However, there are few literatures on the relationship between farmland property right and irrigation collective action.

3.2.4 Logical Relationship between Farmland Resources and Rural Collective Action based on IAD Framework

By analyzing the relationship between the farmland resources and rural irrigation collective action, it is known that farmland resources have different effect on rural irrigation collective action because of the characteristics of the inherent in farmland resources. In order to further analyze the logical relationship between farmland resources and rural irrigation collective action, IAD is still used to make further explanation in this book (see Figure 3-2).

The different attributes of farmland resources affect the rural irrigation collective action from different angles. For example, the scale of farmland

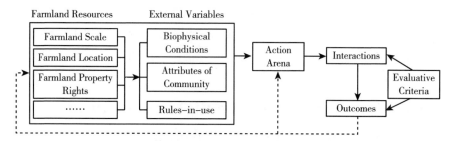

Figure 3-2 Logical relationship between farmland resources and rural collective action based on IAD framework

can be ascribed to the biophysical conditions of the external variables of collective action, but at the same time, the formation of the scale of farmland can also be attributed to the different arrangements of the property right distribution system in addition to the natural factors, so it can also be summarized to the attribute of the rules-in-use of the external variables of collective action. Similarly, although the location of farmland can be classified as an attribute of biophysical conditions, it also has the influence of different institutional arrangements, so it can also be summed up as the influence of general institutional rules. In conclusion, among the external variables of collective action, farmland resources can be classified into different types, which can exert important driving or inhibiting effects on the formation of rural irrigation collective action from different perspectives.

3.3 Logical Relationship between Farmland Fragmentation and Rural Irrigation Collective Action

The farmland resources held by a small-scale peasant household may be divided into multiple plots due to terrain and other conditions, or it may be divided into multiple plots due to the fairness of property rights distribution. In other words, it has the characteristics of natural fragmentation and property rights fragmentation. Especially for China, due to the large rural population, the property rights of farmland are constantly adjusted at the village level based on the principle of fairness according to the natural characteristics such as the size or the distance of the land. Therefore, the farmland fragmentation has become another important characteristic of small-scale peasant households in China. The fragmentation of natural and property rights affects the rural collective action both from the biophysical conditions and the rules-in-use in the external variables of collective action.

Small-scale peasant households with different degrees of farmland fragmentation usually show behavioral preferences and demand preferences with obvious individual characteristics, and their demands for rural production public goods also present different preferences. As a result, small-scale peasant households with different degrees of farmland fragmentation also have great differences in their enthusiasm and choice of ways to engage in rural collective action (Lyu, 2011). The characteristics of farmland fragmentation of small-scale peasant households may also influence the formation of rural collective action from both positive and negative aspects. First of all, the characteristics of farmland fragmentation unique to small-scale peasant households in China may directly reduce the capacity of rural collective action, that is, the smaller the area of a single plot of farmland in a household and the larger the number of plots, the worse the degree of small-scale peasant households' participation in collective action, that is, the farmland fragmentation has a negative impact on the participation of small-scale peasant households in collective action. The reason is that the characteristics of small-scale peasant households' farmland fragmentation may affect the formation of rural collective action in many aspects. For example, farmland fragmentation may weaken rural collective action capacity by reducing small-scale peasant households' dependence on agricultural production, increasing the pressure on peasants to engage in rural collective action, and increasing the differences among different small-scale peasant households (Manjunatha, 2013). Secondly, the characteristics of farmland fragmentation of small-scale peasant households may also enhance the collective action capacity to some extent. For example, the land fragmentation of small-scale peasant households promotes the implementation of land transfer behavior, and scale management can promote the success of rural collective action to a certain extent (Araral, 2009; Sharaunga, 2017).

3.4 Conclusion

The development and change of small-scale peasant households are closely related to the formation of rural irrigation collective action. As an important input to agricultural production, small-scale peasant households' farmland resources to a large extent affect whether rural irrigation collective action can be formed. The farmland resource attributes such as the scale of farmland, the location of farmland and the property rights of farmland have influences on the rural irrigation collective action from different angles, which have positive promoting effect and negative inhibiting effect. From the perspective of external influence factors of Institutional Analysis and Development framework, these attributes can also be classified into different external variables, which exist in biophysical conditions, attributes of community and rules-in-use, and the combined action of these factors contributes to or reduces rural irrigation collective action from multiple aspects. However, China has a large population, insufficient per capita

CHAPTER 3
LOGICAL RELATIONSHIP BETWEEN FARMLAND RESOURCES AND RURAL IRRIGATION COLLECTIVE ACTION

resource endowment, and the phenomenon of farmland fragmentation is serious. The research on how to judge the ultimate influence of the special phenomenon of farmland fragmentation on the rural irrigation collective action is of great significance not only to enrich the existing collective action research framework, but also to the concrete practice. This chapter only discusses the logical relationship between farmland fragmentation and rural irrigation collective action from the theoretical part. In order to further judge the degree of interaction between the two, the following chapters use econometric analysis method to discuss.

CHAPTER 4

IMPACT OF FARMLAND RESOURCE ENDOWMENT ON RURAL IRRIGATION COLLECTIVE ACTION

In order to explore the impact of farmland fragmentation on rural irrigation collective action, this book first analyzes the impact of farmland resource endowment on rural irrigation collective action from a macro perspective, so as to preliminarily judge the role played by farmland resource endowment in the formation of rural irrigation collective action in China. Meanwhile, it will also pave the way for the subsequent analysis of the impact degree and specific impact mechanism of farmland fragmentation. This chapter first describes the data collection status used in this study, then constructs the model of collective action analysis of the impact of farmland resource endowment on rural irrigation, and finally draws research conclusions through econometric analysis.

CHAPTER 4
IMPACT OF FARMLAND RESOURCE ENDOWMENT ON RURAL IRRIGATION COLLECTIVE ACTION

4.1 Data

The data used in this book are mainly from the survey data of China Institute for Rural Studies, Tsinghua University in 2017. The main purpose of the survey is to grasp the basic situation of the country's rural areas, including agricultural production, farmers' income and expenditure, rural environmental conditions, agricultural and rural policy implementation, road and irrigation and other rural infrastructure construction and maintenance conditions.

The survey adopted questionnaire, including village-level questionnaire and household-level questionnaire. The village-level questionnaire is mainly filled out by leaders familiar with the situation of the village, such as the village head or village director, accounting personnel, etc. The contents of the questionnaire mainly include the basic situation of the administrative village, land, homelands and housing, infrastructure and public services, village-level economy and income and expenditure, irrigation and water environment conditions, labor transfer and employment, village governance conditions, etc. The household-level questionnaire is mainly filled out by household heads randomly selected by scholars. The contents of the questionnaire mainly include basic information of family members, infrastructure and living environment, land, homelands and houses, water conservancy and irrigation, consumption and income, participation in village governance, etc.

From April to May 2017, China Institute for Rural Studies, Tsinghua University recruited students from universities across the country. In the end, more than 900 students were recruited, mainly from agricultural universities across the country, such as China Agricultural University,

Sichuan Agricultural University, Hebei Agricultural University. In June 2017, four teachers and experts from China Institute for Rural Studies, Tsinghua University gave training to the students who engaged in the survey. The training content mainly included the answer to the questionnaire and the selection method of peasant household samples.

The survey was conducted from June to September 2017, during the summer vacation of various universities. Considering the convenience and cost of research, the vast majority of scholars were required to conduct research in their hometown or school location. Finally, 865 village-level questionnaires and 17,949 household-level questionnaires were collected. The purpose of this chapter is to analyze the influence of farmland endowment on rural irrigation collective action from the village level. Therefore, the analysis in this part only selects data at the village-level for quantitative analysis. According to the availability of data used in this chapter, a total of 283 village-level data were selected to be included in the econometric analysis.

4.2 Model

There are many indicators to measure the rural irrigation collective action, and the research angles are different. For example, from the perspective of peasants' current behaviors, these indicators include peasants' participation in the maintenance of rural irrigation facilities, investment status, cost sharing status, etc. From the perspective of future behavior, these indicators include the willingness to reform the collective water system, the willingness to pay for water in the future, and the willingness to invest in facility construction. However, since this part studies the rural

CHAPTER 4
IMPACT OF FARMLAND RESOURCE ENDOWMENT ON RURAL IRRIGATION COLLECTIVE ACTION

irrigation collective action from the village level, the indicator of how the village collective irrigation facilities are maintained in the past three years is selected as the village-level indicator to measure the rural irrigation collective action in combination with the specific questionnaire in this study. In the process of specific survey, as the questionnaire was set up using Likert scale, ordered Probit model was planned to be used for analysis in the measurement research of this chapter. In terms of the endowment of farmland resources, this study takes the area of farmland resources per capita of the population in the village as a specific measurement indicator.

In addition to the above indicators, this chapter also select other variables as control variables, such as the terrain of the village, the water resources of the village, the distance between the village and the city, whether the village is dominated by planting industry, whether it is a poor village, whether there are punishment measures for stealing water at the village level, the failure of village governance, whether there is a water association in the village, and government fund support.

The econometric model (4-1) is constructed in this chapter:

$$Y_i^* = \beta X + \varepsilon, i = 1,2,\cdots,5 \qquad (4-1)$$

The evaluation indicators of rural irrigation collective action can be grouped into five different levels. Its essence is a sorting and selection problem. Therefore, this study adopts the ordered Probit model to analyze the influence of farmland resource endowment on farmers' participation in irrigation collective action, with the specific formula as follows:

$$Y_i^* = \beta X + \varepsilon, i = 1,2,\cdots,5 \qquad (4-2)$$

$$Y_i = \begin{cases} 1, & \text{if } Y_i^* \leq \alpha_1 \\ 2, & \text{if } \alpha_1 < Y_i^* \leq \alpha_2 \\ \cdots\cdots \\ 5, & \text{if } Y_i^* > \alpha_4 \end{cases} \qquad (4-3)$$

The probabilities of $Y_i = 1, 2, \cdots, 5$ are:

$$\begin{cases} Prob(Y_i = 1|X) = Prob(\beta X + \varepsilon \leqslant \alpha_1 |X) = \varphi(\alpha_1 - \beta X) \\ Prob(Y_i = 2|X) = Prob(\alpha_1 < \beta X + \varepsilon \leqslant \alpha_2 |X) = \varphi(\alpha_2 - \beta X) - \varphi(\alpha_1 - \beta X) \\ \cdots \cdots \\ Prob(Y_i = 5|X) = Prob(\beta X + \varepsilon > \alpha_4 | X) = 1 - \varphi(\alpha_4 - \beta X) \end{cases}$$

$$(4-4)$$

Y_i^* is a potential variable and can't be observed, but Y_i is an observable variable; X is a set of observed values of explanatory variables including farmland resource endowment variables; β represents the parameter variable to be estimated; ε denotes residual; α denotes the interval demarcation point; φ denotes the standard normal cumulative distribution function.

4.3 Results

4.3.1 Descriptive Analysis

By means of the econometric analysis model constructed above, this book uses Stata software for analysis. Firstly, descriptive analysis was performed on the dependent variables and independent variables. Table 4–1 shows the results for each indicator. As can be seen from the table, in terms of collective action on rural irrigation, the average value of the index of how the village collective irrigation facilities are maintained in the past three years is 3.739, that is, the current maintenance status of collective irrigation facilities in most villages is at an average level. The maximum per capita farmland resource area is 1.609 ha, the minimum is 0.006 ha, and the average is only 0.213 ha. It indicates that the per capita farmland resource area of the surveyed villages is small.

CHAPTER 4
IMPACT OF FARMLAND RESOURCE ENDOWMENT ON RURAL IRRIGATION COLLECTIVE ACTION

Table 4-1   Descriptive results of village-level variables

Variables	Mean value	Min. value	Max value
Rural irrigation collective action (How the village collective irrigation facilities are maintained in the past three years)	3.739	1	5
Area of farmland resources per capita	0.213	0.006	1.609
Village terrain	0.488	0	1
Village water resources	0.177	0	1
Distance between the village and the city	2.823	0	4.7
Whether the village is dominated by planting industry	0.820	0	1
Whether it is an official certified poor village	0.191	0	1
Whether there are punishment measures for stealing water resources	0.251	0	1
Failure of village governance	1.558	1	5
Whether there is a water association	0.261	0	1
Government fund support	0.541	0	1

4.3.2 Results of Econometric Analysis

Table 4-2 shows the results of the econometric analysis using the Stata software. The values in the table are the coefficient values of the regression results. Model 1 is the measurement result without adding the control variable of each province, and Model 2 is the measurement result with adding the control variable of each province.

Table 4-2 Farmland resource endowment influences the measurement results of rural irrigation collective action

Variables	Model 1	Model 2
Core independent variable		
Endowment of farmland resources	0.780*	0.772*
Control variables		
Village terrain	0.125	0.301*
Village water resources	-0.383**	-0.181

continued

Variables	Model 1	Model 2
Distance between the village and the city	-0.146*	-0.064
Whether the village is dominated by planting industry	-0.259	-0.075
Whether it is an official certified poor village	-0.281	-0.351*
Whether there are punishment measures	-0.201	-0.086
Failure of village governance	-0.111*	-0.082
Whether there is a water association	0.260	0.359
Government fund support	-0.091	0.013
Province	No	Yes
Sample	283	283
Chi2	22.23	77.95
r2_p	0.0308	0.1081

Note: *** means that the results are significant at the level of 1%, ** means that the results are significant at the level of 5%, and * means that the results are significant at the level of 10%.

Table 4-2 shows that the independent variable of farmland resource endowment presents a positive effect in both models, and it is significant at the level of 10%. The econometric results show that the variable of farmland resource endowment has a significant positive promoting effect on the formation of rural irrigation collective action, that is, the larger the per capita area of farmland resources is, the stronger the farmer's participation in rural irrigation collective action will be, and the better the maintenance effect of the collective irrigation facilities in the whole village will be. This fully indicates that the larger the area of farmland resources held by the peasants in a certain region (village), the more the peasants in the whole village expect to be able to rely on agriculture for living. Therefore, in order to maintain the smooth progress of agricultural production activities, peasants are more inclined to maintain the rural public goods (irrigation facilities) through collective action, so as to reduce the high cost of agricultural production through collective action, and thus contribute to the success of rural irrigation collective action.

The above research results are consistent with the research results of other scholars in China. For example, Mao Miankui (2016) used the data from 3,552 surveys of farmers in 21 provinces to find that the larger the contracted farmland area, the stronger the willingness of farmers to participate in the construction and management of small irrigation and water conservancy. Yuan Junlin (2016) analyzed 192 sample farmers in Henan Province and found that the larger the area of farmland owned by peasants, the stronger the degree of participating in the management and protection of small irrigation and water conservancy facilities. Cai Jingjing (2015) obtained the above results after analyzing 430 household survey data from 15 provinces in China. It can be seen that although per capita farmland resources are relatively small in rural areas of China, the higher the degree of farmland resource endowment, the stronger the degree and willingness of farmers to engage in rural irrigation collective action. Peasants will maintain sustainable agricultural production through collective cooperation, and thus improve the peasants' income by reducing the maintenance cost of irrigation facilities. Therefore, from the perspective of the whole village, in terms of rural irrigation collective action, the stronger the farmland resource endowment in the village, the higher the collective action capacity of the whole village on rural irrigation.

4.4 Conclusion

This chapter mainly discusses the impact of farmland resource endowment on rural irrigation collective action from the perspective of village level. By taking how the village collective irrigation facilities are maintained in the past three years as the measuring variable of rural irrigation collective action, and the area of farmland resources per capita of the population in the village as the measuring variable of farmland resource endowment, the ordered Probit model was used to measure the impact of farmland resource endowment on rural irrigation collective action. The results show that although the per capita resources in rural areas of China are small, the higher the degree of farmland resource endowment, the better the performance of the collective action of irrigation in the whole countryside. The research conclusions obtained from the macro perspective in this chapter, on the one hand, lay a good cognitive foundation for the follow-up research on the micro level, on the other hand, are in line with the research results of current scholars to a large extent, indicating that the data used in this study have a certain credibility, and provide better data support for the follow-up research.

CHAPTER 5

TOTAL EFFECT OF FARMLAND FRAGMENTATION ON RURAL IRRIGATION COLLECTIVE ACTION

Fragmentation is one of the unique properties of China's rural farmland. Does it have an impact on the rural irrigation collective action? If yes, what factors can mediate between the two? Based on these two questions, this chapter uses field survey data and builds an econometric model to study the total effect of farmland fragmentation on rural irrigation collective action through empirical analysis. The structure of this chapter is as follows: first, construct the theoretical framework of studying the relationship between farmland fragmentation and rural irrigation collective action based on the IAD framework; second, construct the indicator system and econometric analysis model; finally, analyze and discuss the empirical results.

5.1 Framework

5.1.1 IAD Framework

The theory of collective action points out that individual rationality and collective rationality in collective action are not exactly the same, and the phenomenon of free-riding in which individuals participate in the supply of public goods is common. According to Elinor Ostrom, collective action is not only affected by the characteristics of the actors and the trust relationship between the actors, but also by the resource factors, institutional characteristics and other aspects. To this end, Elinor Ostrom divided the external factors that affect collective action into three types—biophysical conditions, attributes of community and rules-in-use—and established the IAD framework (Ostrom, 2005). Besides, Elinor Ostrom pointed out that these three types of factors do not all directly affect collective action, but encourage users to adjust their participation behaviors by changing the incentive structure they face, and finally affect collective action.

The resources held by the individual user have a great influence on his or her decision-making behavior because the individual user has to make decisions according to the objective realistic conditions (Doss, 2015). As the most basic element of agricultural production, farmland has a particularly important impact on peasant household's production and management or life decisions. Especially in China, more people and less land is the reality of China's national conditions and agricultural conditions, and the impact of land on production and management is particularly prominent. For the object of this book, in the process of peasant households' participation in village collective action, their participation behavior needs be based on the existing farmland conditions. Therefore, the size, quantity, distance, shape and other characteristics of farmland play a role in promoting or restricting peasant households' participation in collective action.

Under the background of household contract as the basic system of production and management, inheritance and other external factors lead to the obvious enhancement of stratified evolution caused by farmland fragmentation (Ye, 2008). Peasant households with different degrees of farmland fragmentation show more obvious behavior preferences and demand preferences with individual characteristics. Their demands for irrigation facilities and other rural public goods also present different preferences, resulting in the peasants with different degree of farmland fragmentation participating in the supply of rural public goods with different enthusiasm and choice of ways.

Based on the above analysis, an analysis framework based on the IAD framework is constructed in this book (see Figure 5-1).

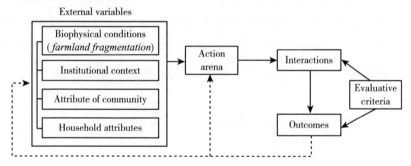

Figure 5-1 Logical relationship between farmland fragmentation and rural irrigation collective action based on IAD framework

5.1.2 Selection of Influencing Factors and Research Hypothesis

5.1.2.1 Farmland Fragmentation

According to the existing studies, there are two characteristics of fragmented farmland: (1) a single peasant household owns multiple plots of non-adjacent farmland; (2) the average area of the plot is small, that is, a single piece of farmland has a small area. However, in addition to studying the direct impact of farmland fragmentation on rural irrigation collective action, this chapter also adopts the cross-term approach to further explore the regulatory effect of the attributes of institutional rule on the relationship between farmland fragmentation and rural irrigation collective action. Therefore, only the first feature of farmland fragmentation is selected as the measurement indicator in this chapter, that is, the number of farmland plots is selected as the specific indicator.

Under normal circumstances, if two peasant households own the same total area of farmland, the peasant household with a larger number of plots needs to spend more time, labor, capital and other inputs on the construction and maintenance of irrigation facilities. Therefore, the degree of participation in irrigation collective action is lower than that of peasant households with smaller plots. To this end, this part assumes that farmland fragmentation has a negative impact on the rural irrigation collective action.

5.1.2.2 Attributes of Institutional Context

(1) Stability of property rights

Under the current land system, every member of a village in China enjoys equal rights to contract and manage farmland. Therefore, with the change of the population in the village, the village will adjust the land property rights periodically or irregularly. Irrigation facilities have fixed characteristics in geographical location, and long-term characteristics in use. Since the construction and maintenance of irrigation facilities require a large amount of funds, farmers may choose to make short-term investments in order to ensure agricultural output if the property rights of farmland are unstable, so they may not engage in irrigation collective action.

(2) Irrigation operational rule

Whether there is a unified irrigation management system is an important factor affecting the formation of rural irrigation collective action. In general, there are two main irrigation management modes: collective management and individual management. Compared with the two management modes, collective management has certain institutional rules. Especially in the areas with poor natural conditions of water resources, collective management can better constrain the irrigation behavior of peasant households and thus maintain the fairness of each member in the village. Therefore, collective management is generally able to form unified institutional rules and may encourage peasants to engage in rural irrigation collective action.

(3) Failure of village governance

If there are many water disputes among peasants, it means that peasants cannot trust each other, that is, village governance fails. Therefore, peasants will be less willing to participate in the construction, maintenance and decision-making of rural irrigation facilities, and irrigation collective action is difficult to form.

5.1.2.3 Biophysical Conditions

(1) Distance to city

If the village is closer to the city, peasants have more opportunities to work in the city, that is, there are more part-time workers in the villages closer to the city, and they are less dependent on agricultural production, so they are less willing to engage in the rural irrigation collective action.

(2) Village water resource

If the natural conditions of water resources in the region where the village is located are good, the farmland can be completely dependent on natural resources such as rainwater for irrigation, so the willingness to take collective action will be weak. If water resources are poor, objective and realistic conditions have determined that the construction and maintenance of irrigation facilities will achieve poor benefits, and peasants will be less willing to participate in collective action. Therefore, when water resources conditions are moderate, peasants may choose to take collective action.

(3) Farmland size

The scale of farmland affects the enthusiasm of peasants to engage in the rural irrigation collective action to a large extent. In general, if peasants own a large area of farmland, they will have a higher demand for irrigation facilities, so they will have a stronger willingness to carry out construction and maintenance of irrigation facilities.

(4) Farmland location

According to the existing literature, scholars believe that the location of farmland has a significant impact on peasants' participation in irrigation collective action. If the farmland is located in the upstream of the irrigation water source, the peasants will be less willing to participate in the construction and maintenance of collective irrigation facilities due to the relatively abundant water resources. If it is located downstream of irrigation water source, the water resource is poor, and the income brought by more input is not obvious, and there is diminishing marginal benefit. Therefore, peasants will be less willing to engage in irrigation collective action.

(5) Village topography

There are two possible effects of village topography on peasants' participation in collective irrigation actions. First, if the village is located in a plain, the cost of constructing and maintaining irrigation facilities such as water channels will be lower, and peasants are more willing to participate in collective irrigation. Second, irrigation facilities are the infrastructure of agricultural production. If the village is located in mountains and hills, the cost of individual construction of irrigation facilities will be higher. Therefore, peasants may reduce the cost of construction and maintenance of irrigation facilities through collective action.

(6) Farmland water scarcity

In general, when water resources are abundant in the location of family farmland, there is less water shortage of farmland, and peasants are less likely to engage in irrigation collective action. When water resources are scarce, peasants are more likely to engage in collective action on irrigation.

5.1.2.4 Attributes of Community

(1) Village size

Olson (1965) believed that with the increase of the number of members in a group, the free-riding dilemma in collective action would also be enhanced. In the villages with a large number of peasant households, each family can get a smaller share of income from irrigation collective action, it is difficult for families to supervise each other, and the cost of organizing irrigation collective action is higher. Therefore, irrigation collective action is more easily formed in smaller villages than in larger villages.

(2) Village economic development

Capital input is the most basic input for the construction and maintenance of irrigation infrastructure. If the village economy is highly developed, the funds for construction and maintenance are guaranteed. Therefore, peasant households in villages with higher economic development are more willing to engage in irrigation collective action.

5.1.2.5 Household Attributes

(1) Family size

The number of households may have two opposite effects on rural irrigation collective action. First, the large number of families has enough labor input in agricultural production and the construction and maintenance of irrigation facilities, so the willingness to engage in irrigation collective action will be stronger. Second, due to the large number of families, the labor force may choose to go out for work. In agriculture, the labor input in production and the construction and maintenance of irrigation facilities is insufficient, and the collective action capacity of irrigation is decreased.

(2) Age of family head

At present, the phenomenon of "hollow village" has become a common phenomenon in China's rural areas, that is, the older peasants are engaged in agricultural production activities, while young people often choose to go out for work. When taking part in irrigation collective action, because the construction and maintenance of irrigation facilities require physical labor, older peasants are often reluctant to take part in collective action due to physical strength and other reasons.

(3) Education of family head

Family head with a higher level of education tends to have a long-term vision and market awareness, and are able to fully realize the importance of the construction and maintenance of irrigation facilities for agricultural production. Therefore, they are more likely to engage in the rural irrigation collective action (Murtinho, 2016).

(4) Farmland ownership transfer

There are two main ways in which land ownership transfer affects irrigation collective action. On the one hand, the land ownership transfer makes some people no longer engage in agricultural production, and the cooperative scale of irrigation collective action is gradually reduced, and the cooperative cost is reduced, making it easier to achieve irrigation collective action. On the other hand, with the land ownership transfer, the scale of production and operation of farmers who rely on planting industry in villages expands, and these groups have stronger demands for collective irrigation, so it is easier to contribute to the success of collective action.

(5) Importance of crop income

The higher the proportion of the income from planting industry in the total household income, the stronger the dependence of the family on planting industry, and the peasants are more inclined to construct and maintain agricultural irrigation facilities, so the peasants are more likely to engage in the rural irrigation collective action.

5.2 Sample Selection and Method

5.2.1 Sample Selection

Since this book only involves the rural areas where canal irrigation is the main irrigation method, only those villages related to canal irrigation are selected in this book out of 865 villages surveyed. In the end, a total of 284 villages have been selected in this chapter, involving 3,895 peasant households in 17 provinces.

5.2.2 Method

By combing the current literature on collective action, it is found that there are two main methods for scholars to evaluate collective action—output method and process method. Output method refers to the result of collective action to measure collective action, while process method refers to the process of collective action to measure collective action. To evaluate the rural irrigation collective action objectively, the output method and process methods are used to measure the rural irrigation collective action. Therefore, the output method is used to select peasants' participation in the maintenance of collective water conservancy project facilities as the evaluation indicator of irrigation collective action, and the process method is used to select peasants' participation in the decision-making of collective water conservancy project meetings as the evaluation indicator of irrigation collective action. Then, the ordered Probit model is used for analysis. According to the above theoretical analysis framework and econometric model, the names, meanings and expected directions of variables used in this book are shown in Table 5–1.

Table 5–1   Variables definition and expected impact on irrigation collective action

Variables	Definition and measurement	Expected sign
Dependent variables		
CLTMTN	Frequency of participation in collective maintenance: 1 = never, 2 = seldom, 3 = sometimes, 4 = often and 5 = usually	
CLTMET	Frequency of attending village meetings related to irrigation: 1 = never, 2 = seldom, 3 = sometimes, 4 = often and 5 = usually	
Independent variables		
Farmland fragmentation		
Plot number	The number of farmland plot in a household	-
Institutional context		
Stability of property rights	1 if there was no massive adjustment of land tenure in the village over the past years, 0 otherwise	+
Irrigation operational rule	1 if there is an irrigation operational rule in the village; 0 otherwise	+
Failure of village governance	1 if farmer think there are too many conflicts in the process of canal utilization; 0 otherwise	-

continued

Variables	Definition and measurement	Expected sign
Biophysical conditions		
Distance to city	The distance from village to the nearest city (km)	+
Village water resource	1 if there is moderate water scarcity at village level; 0 otherwise	+
Farmland size	The total area of farmland owned by household (ha)	+
Farmland location	1 if the farmland is located on the midstream of the canal; 0 otherwise	+
Village topography	1 if the village is located in a plain area; 0 otherwise	+/−
Farmland water scarcity	The degree of farmland water scarcity in a household: 1 = never, 2 = not serious, 3 = normal, 4 = serious and 5 = extremely serious	+
Attributes of community		
Village size	The total households in a village	−
Village economic development	The total income of a village in 2016 (US $)	+
Household attributes		
Family size	The number of members in a family	+/−
Age of family head	The age of the household head	−
Education of family head	The education of household head: 1 = no school, 2 = primary school, 3 = junior high school, 4 = senior high school, 5 = college, 6 = graduate	+
Farmland ownership transfer	1 if the household transfer in or out of the farmland; 0 otherwise	+
Importance of crop income	1 if the income mainly comes from crop in a household total income; 0 otherwise	+

5.3 Results

5.3.1 Descriptive Analysis

Table 5-2 shows the statistical results of each indicator. As can be seen from the table, the mean of the indicator the frequency of participation in collective maintenance is 3.201, while the mean of the indicator the frequency of attending village meetings related to irrigation is 3.163. In terms of farmland fragmentation, the plot number ranges between 1 and 28, with an average of 4.127. It fully showed that each family has a lot of farmland plots i.e. farmland fragmentation is very serious in China.

There is a short distance from village to city because there is an average of 22.112 km among 284 villages, ranging between 1 and 105 km. As for the farmland size, it ranges between 0.010 ha and 17.070 ha, with an average of 0.258 ha. There is a large gap of village size among 284 villages because it ranges between 25 and 4,687 households and the standard deviation is 643.512, with an average of 720.552 households. Likewise, the gap of village development is similar with village size, because it ranges between 0 and 7.528 million dollars, with an average of 0.142 million dollars and a standard deviation of 80.944. In terms of household attributes, family size is between 1 and 10 and each family has about 3 members on average. The age of household head is between 20 and 105 years old with an average of about 52 years old. The average educational level of the household head is 2.683, which means that the education level of most household heads are the level of primary or middle school. The mean of tenure stability of property rights, irrigation operational rule, failure of village governance, village water resource, village topography, farmland location, farmland water scarcity, farmland ownership transfer, and importance of crop income are 0.866, 0.688, 0.064, 0.160, 0.503, 0.219, 2.905, 0.242 and 0.569, respectively.

Table 5-2   Descriptive statistics

Variables	Mean	Std. Dev.	Min.	Max
CLTMTN	3.201	1.086	1	5
CLTMET	3.163	1.056	1	5
Plot number	4.127	2.922	1	28
Stability of property rights	0.866	0.340	0	1
Irrigation operational rule	0.688	0.463	0	1
Failure of village governance	0.064	0.244	0	1
Distance to city (km)	22.112	15.213	1	105
Ln(Distance to city)	2.862	0.735	0	4.654
Village water resource	0.160	0.367	0	1
Farmland size (ha)	0.258	0.392	0.010	17.070
Ln(Farmland size)	-1.676	0.787	-5.010	2.840
Village topography	0.503	0.500	0	1
Farmland location	0.219	0.413	0	1
Farmland water scarcity	2.905	1.159	1	5
Village size (household)	720.522	643.512	25	4687
Ln(Village size)	6.255	0.824	3.219	8.453
Village economic development (million dollars)	0.142	80.944	0	7.528
Ln(Village development)	-5.863	4.491	-8.800	6.620
Family size	3.206	1.355	1	10
Age of family head	52.096	12.918	20	105
Education of family head	2.683	0.940	1	6
Farmland ownership transfer	0.242	0.428	0	1
Importance of crop income	0.569	0.495	0	1

5.3.2 Results of Econometric Analysis

To understand the influence of farmland fragmentation and other variables on irrigation collective action, the author employ stepwise regression. Table 5-3 and Table 5-4 show the results. In these two tables, Model 1 and Model 8 only added the variables of farmland fragmentation; Model 2 and Model 9 added the variables of institutional context; Model 3 and

Model 10 added the variables of physical attributes; Model 4 and Model 11 added the variables of community attributes; Model 5 and Model 12 added the variables of household attributes; Model 6 and Model 13 added the regional control variables. Furthermore, Model 7 and Model 14 were obtained using the least square method. From the results of seven models in each table, the sign of most independent variables has not substantially changed after adding different control variables. It indicates that these models are robust.

Table 5-3   The determinants of participation in construction and maintenance of collective irrigation

		Model 1	Model 2	Model 3	Model 4	Model 5	Model 6	Model 7
Farmland fragmentation	Plot number	0.011 *	-0.013 **	-0.018 ***	-0.015 **	-0.017 **	-0.014 **	-0.014 **
		(0.006)	(0.006)	(0.006)	(0.006)	(0.007)	(0.007)	(0.007)
Institutional context	Stability of property rights		0.179 ***	0.138 ***	0.155 ***	0.155 ***	0.233 ***	0.229 ***
			(0.05)	(0.051)	(0.051)	(0.051)	(0.055)	(0.054)
	Irrigation operational rule		0.145 ***	0.132 ***	0.137 ***	0.129 ***	0.039	0.034
			(0.037)	(0.040)	(0.040)	(0.040)	(0.042)	(0.040)
	Village governance failure		-0.747 ***	-0.734 ***	-0.720 ***	-0.704 ***	-0.880 ***	-0.858 ***
			(0.07)	(0.073)	(0.073)	(0.078)	(0.082)	(0.079)
Physical attributes	Distance to city			0.089 ***	0.079 ***	0.073 ***	-0.037	-0.035
				(0.024)	(0.025)	(0.025)	(0.028)	(0.027)
	Village water resource			-0.014	0.006	0.001	0.062	0.058
				(0.048)	(0.049)	(0.049)	(0.053)	(0.051)
	Farmland size			0.041 *	0.029	0.014	0.001	0.003
				(0.024)	(0.024)	(0.025)	(0.027)	(0.026)
	Village topography			-0.006	0.002	-0.009	-0.041	-0.038
				(0.039)	(0.039)	(0.039)	(0.043)	(0.041)
	Farmland location			0.064	0.058	0.052	0.045	0.042
				(0.041)	(0.041)	(0.041)	(0.042)	(0.041)
	Farmland water scarcity			0.144 ***	0.142 ***	0.136 ***	0.139 ***	0.132 ***
				(0.015)	(0.015)	(0.015)	(0.016)	(0.015)
Community attributes	Village size				-0.069 ***	-0.063 ***	-0.07 **	-0.066
					(0.022)	(0.023)	(0.030)	(0.029)
	Village development				0.007 *	0.008 *	0.001	0.001
					(0.004)	(0.004)	(0.005)	(0.005)

continued

		Model 1	Model 2	Model 3	Model 4	Model 5	Model 6	Model 7
Household attributes	Family size					0.015 (0.013)	0.020 (0.014)	0.019 (0.013)
	Age of family head					-0.003** (0.001)	-0.003** (0.001)	-0.003** (0.001)
	Education of family head					0.013 (0.021)	-0.001 (0.022)	-1.625E-04 (0.021)
	Farmland ownership transfer					0.103** (0.041)	0.078* (0.041)	0.074* (0.040)
	Importance of crop income					0.158*** (0.036)	0.136*** (0.037)	0.133*** (0.036)
	Province	no	no	no	no	no	yes	yes
	Sample	3895	3895	3895	3895	3895	3895	3895
	Chi2	3.77	139.60	256.98	268.06	300.15	501.46	
	r2_p	0.0003	0.0122	0.0225	0.0235	0.0263	0.0439	
	R-squared							0.1213

Notes: ***, ** and * indicate significant at the 0.5%, 2.5% and 5% level, respectively.

Table 5-4 The determinants of attending meeting of collective irrigation

		Model 8	Model 9	Model 10	Model 11	Model 12	Model 13	Model 14
Farmland fragmentation	Plot number	0.014** (0.006)	-0.014** (0.006)	-0.015** (0.006)	-0.012* (0.006)	-0.013** (0.007)	-0.016** (0.007)	-0.015** (0.007)
Institutional context	Stability of property rights		0.171*** (0.050)	0.111** (0.051)	0.137*** (0.051)	0.137*** (0.051)	0.202*** (0.056)	0.190*** (0.051)
	Irrigation operational rule		0.228*** (0.037)	0.265*** (0.040)	0.268*** (0.040)	0.260*** (0.040)	0.162*** (0.042)	0.145*** (0.039)
	Village governance failure		-0.749*** (0.070)	-0.715*** (0.073)	-0.686*** (0.073)	-0.666*** (0.077)	-0.851*** (0.082)	-0.799*** (0.075)
Physical attributes	Distance to city			0.040* (0.024)	0.030 (0.025)	0.023 (0.025)	-0.084*** (0.028)	-0.076*** (0.026)
	Village water resource			0.056 (0.048)	0.080 (0.049)	0.075 (0.049)	0.127** (0.053)	0.119** (0.049)
	Farmland size			-2.502E-04 (0.024)	-0.015 (0.024)	-0.032 (0.025)	-0.036 (0.027)	-0.031 (0.025)
	Village topography			-0.034 (0.039)	-0.025 (0.039)	-0.036 (0.039)	-0.109** (0.043)	-0.098** (0.040)
	Farmland location			-0.009 (0.041)	-0.020 (0.041)	-0.026 (0.041)	-0.057 (0.042)	-0.054 (0.039)
	Farmland water scarcity			0.173*** (0.015)	0.175*** (0.015)	0.168*** (0.016)	0.160*** (0.016)	0.144*** (0.015)

continued

		Model 8	Model 9	Model 10	Model 11	Model 12	Model 13	Model 14
Community attributes	Village size				-0.100 *** (0.022)	-0.095 *** (0.023)	-0.120 *** (0.030)	-0.108 *** (0.028)
	Village development				0.016 *** (0.004)	0.018 *** (0.004)	0.017 *** (0.005)	0.015 *** (0.004)
Household attributes	Family size					0.018 (0.013)	0.023 * (0.014)	0.020 (0.013)
	Age of family head					-0.003 * (0.001)	-0.003 ** (0.002)	-0.003 ** (0.001)
	Education of family head					0.016 (0.021)	0.002 (0.022)	0.003 (0.020)
	Farmland ownership transfer					0.098 ** (0.041)	0.093 ** (0.041)	0.079 ** (0.038)
	Importance of crop income					0.181 *** (0.036)	0.156 *** (0.038)	0.143 *** (0.035)
	Province	no	no	no	no	no	yes	yes
	Sample	3895	3895	3895	3895	3895	3895	3895
	Chi2	6.27	163.20	305.26	337.54	373.48	649.85	
	r2_p	0.0006	0.0145	0.0271	0.0300	0.0331	0.0577	
	R-squared							0.1542

Notes: ***, ** and * indicate significant at the 0.5%, 2.5% and 5% level, respectively.

This book focused on analyzing the results of Model 6 and Model 13. The farmland fragmentation is an important factor influencing peasants' participation in irrigation collective action, because significant test of 5% level was passed and the coefficient was negative in both models. Namely, if peasants have a lot of farmland plots, they would not like to engage in irrigation collective action, a result consistent with theoretical expectation.

In terms of the variables in the institutional context, both stability of property rights and failure of village governance have passed the significance test in both models while the irrigation operational rule just passed it in the Model 13. From the perspective of impacting direction, the stability of property rights and irrigation operational rule presented a positive impact on collective action and the failure of village governance showed a negative impact. It indicated that peasants would have more enthusiasm to engage in irrigation collective action if the farmland tenure was stable and the village had the operational rule for irrigation. Contrarily, it would inhibit peasants' participation in irrigation collective action if there are more conflicts on utilizing irrigating facilities.

As for the six variables in biophysical conditions, the variable, water scarcity, passed the significance test in both of two models and other three variables, distance to city, village water resource and village topography, passed the significance just in one model and failed significance test in the other model, while the farmland size and farmland location did not pass the significance in both models. In addition, the village water resource and water scarcity presented a positive impact on irrigation collective action and distance to city and village topography presented a negative impact on it. These econometric results fully indicate that if the village's water resource is at a normal level and the farmland owned by a family seriously lack of water, peasants are willing to engage in irrigation collective action. By contraries, if the village is farther away from the urban areas and the village is located in plain area, peasants would not like to engage in the irrigation collective action.

The village size passed a 5% level significance test in Model 6 and a 1% level significance test in Model 13, with the negative coefficients in both models. This result is corresponding to the results of most previous research that the collective action is not easy to be generated if there are more users. Furthermore, the village economic development passed a 1% level significance test in Model 13 and it presented a positive impact on irrigation collective action.

Three variables in household attributes, the age of family head, farmland ownership transfer and importance of crop income, passed the significance test in Model 6 and Model 13, while the family size passed the significance test just in Model 13. From the perspective of impacting direction, the importance of crop income, farmland economic transfer and family size played a positive role in irrigation collective action and the age of family head presented the negative impact.

The above econometric results have fully demonstrated that the farmland fragmentation had a significant and negative impact on irrigation collective action. Therefore, it became a key issue for reducing the negative impact of farmland fragmentation on the collective action in order to improve the ability of collective action in rural area of China. Meanwhile, the above results also showed that reasonable institutional arrangements could contribute to success of irrigation collective action. As such, it is important to explore whether institutional context could affect the relationship between farmland fragmentation and irrigation collective action. This book further analyzed the regulatory role of institutional factors by introducing the interaction of farmland fragmentation with stability of property rights, irrigation operational rule and failure of village governance, respectively.

The econometric results were showed in Table 5-5 and Table 5-6 that the dependent variables from Model 15 to Model 18 are the frequency of participation in collective maintenance, and the dependent variables from Model 19 to Model 22 are the frequency of attending village meetings related to irrigation. Through quantitative analysis, it is found that two interactions, farmland fragmentation multiplied by stability of property rights and irrigation operational rule, are all positive and pass the 1% level significance test, while the interaction of farmland fragmentation multiplied by the failure of village governance is negative but do not pass the significance test.

Therefore, we could make a conclusion that the stable farmland property rights and the irrigation operational rule could significantly reduce the negative impact of the farmland fragmentation on irrigation collective action. Thus, it is important to keep stable land tenure, set down the operational rule for irrigation and improve the village governance in order to further realize the irrigation collective action in China.

Table 5-5 Results of peasants' participation in the construction and maintenance of collective irrigation facilities (with interaction)

		Model 15	Model 16	Model 17	Model 18
Farmland fragmentation	Plot number	-0.083*** (0.017)	-0.046*** (0.010)	-0.014* (0.007)	-0.125*** (0.020)
Institutional context	Stability of property rights	-0.050 (0.086)	0.245*** (0.055)	0.234*** (0.055)	-0.069 (0.086)
	Irrigation operational rule	0.046 (0.042)	-0.201*** (0.072)	0.038 (0.042)	-0.229*** (0.073)
	Village governance failure	-0.881*** (0.082)	-0.899*** (0.082)	-0.73*** (0.176)	-0.672*** (0.176)
Interaction	Plot number × Stability of property rights	0.078*** (0.018)			0.088*** (0.018)
	Plot number × Irrigation operational rule		0.052*** (0.013)		0.059*** (0.013)
	Plot number × Failure of village governance			-0.044 (0.046)	-0.068 (0.046)
Physical attributes	Distance to city	-0.036 (0.028)	-0.036 (0.028)	-0.037 (0.028)	-0.035 (0.028)
	Village water resource	0.057 (0.053)	0.079 (0.053)	0.062 (0.053)	0.077 (0.053)
	Farmland size	-0.001 (0.027)	-0.002 (0.027)	0.001 (0.027)	-0.004 (0.027)
	Village topography	-0.037 (0.043)	-0.019 (0.043)	-0.040 (0.043)	-0.011 (0.043)
	Farmland location	0.049 (0.042)	0.043 (0.042)	0.046 (0.042)	0.046 (0.042)
	Farmland water scarcity	0.140*** (0.016)	0.140*** (0.016)	0.139*** (0.016)	0.140*** (0.016)

continued

		Model 15	Model 16	Model 17	Model 18
Community attributes	Village size	-0.061** (0.030)	-0.075** (0.030)	-0.069** (0.030)	-0.066** (0.030)
	Village economic development	0.001 (0.005)	0.001 (0.005)	0.001 (0.005)	-7.010E-07 (0.005)
Household attributes	Family size	0.021 (0.014)	0.021 (0.014)	0.021 (0.014)	0.022 (0.014)
	Age of family head	-0.003** (0.001)	-0.003** (0.001)	-0.003** (0.002)	-0.003* (0.002)
	Education of family head	-0.001 (0.022)	0.002 (0.022)	-1.004E-04 (0.022)	0.004 (0.022)
	Farmland ownership transfer	0.082** (0.041)	0.085** (0.041)	0.077* (0.041)	0.009** (0.041)
	Importance of crop income	0.135*** (0.037)	0.130*** (0.038)	0.136*** (0.037)	0.128*** (0.038)
	Province	yes	yes	yes	yes
	Sample	3895	3895	3895	3895
	Chi2	519.97	518.07	502.39	542.56
	r2_p	0.0456	0.0454	0.0440	0.0475

Notes: ***, ** and * indicate significant at the 0.5%, 2.5% and 5% level, respectively.

Table 5-6 Results of peasants' participation in collective irrigation facilities related meetings (with interaction)

		Model 19	Model 20	Model 21	Model 22
Farmland fragmentation	Plot number	-0.098*** (0.017)	-0.053*** (0.011)	-0.016** (0.007)	-0.148*** (0.020)
Institutional context	Stability of property rights	-0.139 (0.086)	0.216*** (0.056)	0.202*** (0.056)	-0.159* (0.087)
	Irrigation operational rule	0.171*** (0.042)	-0.121* (0.072)	0.162*** (0.042)	-0.147** (0.073)
	Village governance failure	-0.854*** (0.082)	-0.873*** (0.082)	-0.857*** (0.176)	-0.790*** (0.176)

continued

		Model 19	Model 20	Model 21	Model 22
Interaction	Plot number × Stability of property rights	0.094*** (0.018)			0.104*** (0.018)
	Plot number × Irrigation operational rule		0.061*** (0.013)		0.069*** (0.013)
	Plot number × Failure of village governance			0.002 (0.046)	−0.026 (0.046)
Physical attributes	Distance to city	−0.083*** (0.028)	−0.082*** (0.028)	−0.084*** (0.028)	−0.081*** (0.028)
	Village water resource	0.122** (0.053)	0.148*** (0.053)	0.127** (0.053)	0.145*** (0.053)
	Farmland size	−0.037 (0.027)	−0.039 (0.027)	−0.036 (0.027)	−0.042 (0.027)
	Village topography	−0.104** (0.043)	−0.084* (0.043)	−0.109** (0.043)	−0.075* (0.043)
	Farmland location	−0.053 (0.042)	−0.061 (0.042)	−0.057 (0.042)	−0.057 (0.042)
	Farmland water scarcity	0.160*** (0.016)	0.161*** (0.016)	0.160*** (0.016)	0.162*** (0.016)
Community attributes	Village size	−0.11*** (0.030)	−0.127*** (0.030)	−0.120*** (0.030)	−0.116*** (0.030)
	Village economic development	0.016*** (0.005)	0.016*** (0.005)	0.017*** (0.005)	0.015*** (0.005)
Household attributes	Family size	0.023* (0.014)	0.023* (0.014)	0.023* (0.014)	0.024* (0.014)
	Age of family head	−0.003** (0.002)	−0.003* (0.002)	−0.003** (0.002)	−0.003* (0.002)
	Education of family head	0.002 (0.022)	0.005 (0.022)	0.001 (0.022)	0.006 (0.022)
	Farmland ownership transfer	0.0980 (0.041)	0.101** (0.041)	0.093** (0.041)	0.108*** (0.041)
	Importance of crop income	0.154*** (0.038)	0.149*** (0.038)	0.156*** (0.038)	0.146*** (0.038)
	Province	yes	yes	yes	yes
	Sample	3895	3895	3895	3895
	Chi2	676.50	672.77	649.85	705.04
	r2_p	0.0600	0.0597	0.0577	0.0626

Notes: ***, ** and * indicate significant at the 0.5%, 2.5% and 5% level, respectively.

5.4 Discussion

In general, the smaller the area of individual farmland and the larger the number of plots, the worse the degree of peasants' participation in irrigation collective action, that is, the farmland fragmentation has a negative impact on peasants' participation in irrigation collective action. Through interviews with peasants, it is known that the reasons for the impact of farmland fragmentation on peasants' participation in irrigation collective action are mainly as follows:

First, farmland fragmentation increases agricultural production capital input and raises the total cost of agricultural production for farmers. On the one hand, the farmland fragmentation will enable peasants to make diversified choices for crop planting, which will lead to the increase of input. On the other hand, the farmland fragmentation leads to the waste of agricultural means of production, the increase of supervision and management costs. However, the construction and maintenance of collective irrigation facilities require more funds than other inputs. Therefore, limited input funds, to some extent, limit the construction and maintenance of collective irrigation facilities of farmers, which will lead to the decline of collective action capacity.

Second, the farmland fragmentation increases the labor input in the process of agricultural production and results in the increase of the labor utilization cost. Farmland fragmentation has the characteristics of small area, irregular shape and discrete distribution of a single piece of farmland, and the technologies are difficult to play in the process of agricultural production. Therefore, peasants are more inclined to choose labor input instead of agricultural machinery input. The limited number of labor in the

CHAPTER 5
TOTAL EFFECT OF FARMLAND FRAGMENTATION ON RURAL IRRIGATION COLLECTIVE ACTION

family limits the input of labor in the construction and maintenance of collective irrigation facilities. In addition, the farmland fragmentation increases the time input of labor force, which increases the time spent by peasants traveling to and from different plots for cultivation. The time consumption also limits the participation of peasants in irrigation collective action.

Third, farmland fragmentation reduces the scale effect of agricultural production. Agricultural production is characterized by increasing returns to scale. Irrigation, as the most basic infrastructure of agricultural production, also needs more input. However, farmland fragmentation changes the marginal output elasticity of input factors and reduces the economies of scale in the agricultural production process. Therefore, compared with the households with the characteristics of farmland fragmentation, the peasants' enthusiasm in agricultural input will be lower, including the participation in the construction, maintenance and decision-making of collective irrigation facilities.

Fourth, the farmland fragmentation leads to the shift of labor force to non-agricultural industries, leading to the abandonment of some plots. Compared with the households with a lower degree of farmland fragmentation, the peasant households with a higher degree of farmland fragmentation feature higher capital input, more labor input and weaker scale effect in the process of agricultural production, and the labor return rate of agricultural production is far less than that of non-agricultural industries. Therefore, in households with a higher degree of farmland fragmentation, the labor force is more inclined to shift to non-agricultural industries, which leads to the abandonment of all or part of the land, and the low participation of farmers in agricultural production, so their participation in the construction, maintenance and decision-making of collective irrigation facilities is not strong.

Fifth, the peasant households with higher degree of farmland fragmentation may have the behavior of free-riding. In the process of agricultural production, households with poor farmland fragmentation have a relatively high scale effect, so the input of irrigation facilities will improve agricultural output, and they will more actively participate in the construction, maintenance and decision-making of collective irrigation facilities. Therefore, it is possible for households with a higher degree of farmland fragmentation to use collective irrigation facilities without cost, so there may be a free-riding behavior.

In addition to the above key factors, there are several reasons why farmland fragmentation affects peasants' participation in irrigation collective action. For example, farmland fragmentation leads to more farmland occupied by irrigation facilities. The farmland fragmentation has made it possible for peasants to diversify their planting, and some plots of crops do not need a lot of irrigation. As a result of the farmland fragmentation, peasants are unable to allocate labor and capital reasonably in various plots, so they take a wait-and-see attitude in the construction, maintenance, decision-making and other aspects of collective irrigation facilities.

5.5 Conclusion

Based on the survey data of 3,895 farmers in 284 villages in 17 provinces of China, this chapter studies the impact of farmland fragmentation on rural irrigation collective action. It was found that the degree of farmland fragmentation had a significant effect on the degree of peasants' participation in irrigation collective action. In addition, the study found that the higher the degree of farmland fragmentation, the lower the degree of peasants'

CHAPTER 5
TOTAL EFFECT OF FARMLAND FRAGMENTATION ON RURAL IRRIGATION COLLECTIVE ACTION

participation in irrigation collective action, that is, the greater the number of farmland plots, the lower the degree of peasants' participation in the construction and maintenance of collective irrigation facilities and in decision-making meetings of collective irrigation facilities. At the same time, based on the results of econometrics research and the actual interview records with farmers in the process of data collection, this book believes that there are several reasons why farmland fragmentation affects peasants' participation in irrigation collective action. For example, farmland fragmentation increases the input of capital and labor in the process of agricultural production, reduces the scale effect of agricultural production, makes peasants to give up farming part of the land, and households with a higher degree of farmland fragmentation may have a free-riding behavior.

This chapter also found the influence of other variables on peasants' irrigation collective action. For example, the distance from urban areas, natural conditions of water resources, village economic development degree, irrigation management mode, labor force, water shortage degree of farmland, agricultural dependence and other variables have a positive effect on peasants' participation in irrigation collective action, while village topography, village scale and water disputes have a negative effect. Although this chapter clearly explains the impact of farmland fragmentation on rural irrigation collective action, there are still some shortcomings in this book. For example, the economic and social attributes and institutional rules attributes in this book lack of some key variables, such as village heterogeneity, whether there is penalty system for water use, etc., but the absence of these variables does not affect the credibility of the effect of farmland fragmentation. Based on the research in this chapter, we will continue to discuss the mediating mechanism of farmland fragmentation to reduce the rural irrigation collective action.

CHAPTER 6

MEDIATING EFFECT OF FARMLAND FRAGMENTATION ON RURAL IRRIGATION COLLECTIVE ACTION

In the previous chapter, it is found that farmland fragmentation has a negative impact on the rural irrigation collective action, which shows a strong significance. External influencing factors of rural irrigation collective action are diverse and complex. Although farmland fragmentation has a certain impact on rural irrigation collective action, are there any other factors behind this overall influence effect? If there are mediating factors, what roles do they play between the two? Have they intensified the negative impact of farmland fragmentation on rural irrigation collective action or attenuated the negative effects? Based on these questions, this chapter focuses on the mediating influence mechanism of farmland fragmentation on rural irrigation collective action. Through analysis, it is planned to build a logical relationship diagram between farmland fragmentation and rural irrigation collective action.

The structure of this chapter is as follows: first, establish the potential influence mechanism of farmland fragmentation on rural irrigation collective action through literature mathematics and theories; second, construct the theoretical analysis framework and put forward the research hypothesis; third, introduce the research method and variable setting; finally, analyze the effect of farmland fragmentation on rural irrigation collective action, and construct the logical framework.

6.1 Framework and Potential Mechanisms

6.1.1 SES Framework

Based on the Social-ecological System framework, the specific contents of the four core subsystems of the rural irrigation system are: (1) resource system: irrigation system within the region; (2) resource units: a certain amount of water absorbed by a single peasant household from the irrigation system; (3) governance system: organization, system and rules of irrigation facilities management; and (4) users: peasants. Therefore, the research on rural irrigation collective action can be analyzed based on the SES framework.

6.1.2 Analysis of Potential Mechanism

In order to explore the mediating mechanism of farmland fragmentation affecting rural irrigation collective action, the key lies in judging which elements in the SES framework may be affected by farmland fragmentation from the perspective of the impact effect of farmland fragmentation, and then finding out the possible mediating factors between the two.

Three effects of land fragmentation have been discussed by scholars. The first is its impact on agricultural production, specifically production efficiency, costs, and profits (Hartvigsen, 2014). The second is the impact of land fragmentation on land utilization, including land circulation and sustainable land use (Latruffe, 2014). The third is the impact of land fragmentation on the ecological environment, including biodiversity (Kjelland, 2007). Based on these effects and the specific variables in the SES framework, this book identifies four mediating factors through which land fragmentation might affect irrigation collective action.

First, farmland fragmentation could affect peasants' behavior and willingness to engage in irrigation collective action through dependency on farming for their livelihood. Studies have consistently shown that farmland fragmentation can significantly reduce crop yields (Di Falco, 2010; Rahman, 2009), and thus peasants' income. Profits from agricultural production are smaller than from animal husbandry or handicrafts (Lu, 2017). Smaller profits will prompt peasants to turn to other kinds of work to reduce their dependency on agricultural production (Lu, 2018; Wang, 2016). In general, the path of farmland fragmentation to reduce peasants' dependency on agricultural production is as follows: reduction in crop yields → reducing income → shifting to other non-farm livelihood. And given that irrigation is the basic infrastructure for agricultural production, less dependency on agricultural production means less demand for irrigation. Thus, participation in irrigation-related meetings and maintenance projects— that is, collective action—is reduced (Mushtaq, 2007). Some scholars argue that land fragmentation increases peasants' dependency on farming, because farmland fragmentation can improve the landscape, and thus improve peasants' non-agricultural income through rural tourism (Farley,

2012). But most insist that farmland fragmentation reduces dependency on farming, which is retained as a hypothesis here.

Second, farmland fragmentation could affect irrigation rule-making and therefore peasants' participation in irrigation collective action. More plots and small area are the most significant features of farmland fragmentation. The greater the farmland fragmentation in a given area, the more peasants there will be in that area who want irrigation water. Thus it will be that much harder for them to agree on anything, including water allocation, water price, facility construction, etc (Wang, 2017). Some scholars argue that farmland fragmentation could contribute to the formation of institutional rules, because it implies participation in water user associations, which usually make the rules on water use. But others point out that many water user associations in China are ineffective, despite the rules and regulations they have drawn up (Wang, 2018). Institutional rules are an important influence on irrigation collective action (Ostrom, 2000; Agrawal, 2005). Through regulating peasants' water use, irrigation rules can effectively guarantee the rational allocation of water resources and thus meet the actual needs of water users, which is conducive to irrigation collective action (Araral, 2009; Yahua Wang, 2016). Therefore, irrigation system rule-making may be an important way that farmland fragmentation affects irrigation collective action.

Third, farmland fragmentation could increase the economic pressure on peasants by increasing their production costs, and thus reducing their participation in irrigation collective action. Some scholars argue that farmland fragmentation usually results in more investments in infrastructure and transportation costs (Lu, 2018; Sklenicka, 2017). These costs may increase the economic pressure on peasants (Hua Lu, 2015). More farmland fragmentation may mean higher costs for construction and maintenance of irrigation facilities (Cai, 2016; Cai, 2017). This may reduce peasants' participation in irrigation collective action. However, a few scholars argue that farmland fragmentation could increase peasants' income, because they can choose the crops best suited to the land conditions of each plot (Farley, 2012); that is, peasants with more plots could diversify their planting and have a better chance of maximizing output. Taking all this into consideration, we hypothesize that greater farmland fragmentation will mean less investment in agricultural production, and thus less economic pressure for the peasant to participate in irrigation collective action.

Fourth, farmland fragmentation could affect peasants' participation in irrigation collective action through the buying and selling of land tenure. Farmland fragmentation could encourage peasants to transact land tenure to increase efficiency, because larger plots are usually more efficient (Kjelland, 2007; Lu, 2018). The farmland transaction could influence irrigation collective action in two ways. On the one hand, the peasants who transfer in the land tenure will invest in the public irrigation facilities, because they will need more irrigation (Hua Lu, 2016). On the other hand, if a household with serious farmland fragmentation transfers out the tenure of its plots, they will probably turn to non-agricultural activities. Then, irrigation collective action will not be generated because of the smaller demand for irrigation. But households with serious farmland fragmentation could also optimize their configuration in terms of the labor force, capital, and time inputs into the agriculture. The saved labor, capital and time could be used for the public irrigation facilities, providing great benefits to the non-transferred plots, and thus more willingness to engage in irrigation collective action. Thus, farmland ownership transfer may be an important mechanism through which land fragmentation affects irrigation collective action.

Based on the SES framework and in full combination with the literature review above, this book decomposes the SES framework from the perspective of the impact of farmland fragmentation on rural irrigation collective action (see Table 6-1).

Table 6-1   Decomposition of Social-ecological System framework

First layer	Second layer	Third layer
Social, economic and political settings(S)	—	—
Resource systems (RS)	—	—
Resource units (RU)	RU2- Growth or replacement rate	RU2a-Farmland ownership transfer
Governance systems (GS)	GS5- Operational rules	GS5a-Formal system
Users (U)	U2- Socioeconomic attributes of users	U2a-Economic pressure
	U8- Importance of resource*	U8a-Importance of agriculture
Interaction(I) →Outcomes (O)	I3- Deliberation processes	I3a-Participation in the maintenance of collective water conservancy facilities
	I5- Investment activities	I5a- Participation in decision making of collective water conservancy project meeting
	O1- Social performance measures	O1a-Success of irrigation collective action
Related ecosystems(ECO)	—	—

6.2 Hypotheses

In the real world, there may be many factors affecting irrigation collective action. In this book, guided by our literature review and the CIRS survey of rural areas in China, the author mainly consider the four factors just listed, which could mediate a relationship between land fragmentation and irrigation collective action. This book analyses these mechanisms using the SES framework. Four hypotheses are as follows:

H1: Through dependency on farming, farmland fragmentation has a negative effect on irrigation collective action.

H2: Through formal rule-making, farmland fragmentation has a negative effect on irrigation collective action.

H3: Through economic pressure, farmland fragmentation has a negative effect on irrigation collective action.

H4: Through land ownership transfer, farmland fragmentation has a positive effect on irrigation collective action.

6.3 Method, Variables and Model

6.3.1 Method

The indices of land fragmentation and irrigation collective action are abstract and multidimensional, and there are several potential mediating factors for the effect of land fragmentation on irrigation collective action. Structural equation modeling (SEM) can analyze complex systems through comprehensive use of multiple regression analysis, path analysis, and confirmatory factor analysis.

SEM was first introduced by Jöreskog and Goldberger in the 1970s (Jöreskog, 1972). It was initially applied in psychology and sociology, and later to ecology and the environment (Ülengin, 2010). In recent years, it has been widely used in research in economics and management (Kaltenborn,

2012). SEM can also be called latent variable modelling, because one of its most important advantages is that unobservable variables (latent variables) can be measured by observable variables. By analyzing the multilevel, complex, and causal paths of the irrigation collective action system, we can avoid the errors resulting from traditional statistical methods (Chou, 2002).

The measurement model of latent variables was set as:

$$F = \tau_f + \Lambda_f \cdot LF + \delta \quad (6-1)$$

$$G = \tau_g + \Lambda_g \cdot ICA + \zeta \quad (6-2)$$

Equation (6–1) and equation (6–2) specify the relationship between the exogenous latent variable land fragmentation (LF) and the exogenous observed variable F, and the relationship between the endogenous latent variable irrigation collective action (ICA) and the endogenous observed variable G, respectively. Λ_f is the relationship between the exogenous observed variable and the exogenous latent variable; Λ_g is the relationship between the endogenous observed variable and the endogenous latent variable; δ is the error of the exogenous observed variable F; and ζ is the error of the endogenous observed variable G.

Next, the SEM was set as:

$$ICA = \Gamma \cdot LF + \varepsilon \quad (6-3)$$

where Γ is the coefficient matrix of the exogenous latent variable, land fragmentation, that is, the influence of the exogenous latent variable on the endogenous latent variable, and ε is the residual vector.

6.3.2 Variables

(1) Farmland fragmentation

As mentioned in the previous chapter, the farmland fragmentation mainly includes two characteristics: one is that farmland is divided into multiple plots, and the other is that each plot is too small to be reasonably developed (King, 1982). As the previous chapter discussed the regulatory role of institutional rules in the relationship between farmland fragmentation and rural irrigation collective action, only the variable of plot number is used as the measurement index of farmland fragmentation. In this chapter, the average land plot area is further introduced as the evaluation index.

The total number of plots reflects the unconnected characteristic of land fragmentation. Generally, more plots mean greater fragmentation (Rahman, 2009; Manjunatha, 2013; Ji, 2016). The average plot area reflects the degree of land fragmentation from the perspective of economy of scale: the smaller a plot is, the less chance there is for this kind of economy (Lian, 2014). Since there is an inverse relationship between land area and land fragmentation, we used the reciprocal of the average land plot area as an index of land fragmentation, along with the average number of plots owned by each rural household.

(2) Irrigation collective action

How to measure collective action is the key problem in the empirical study of irrigation collective action. The same as the previous chapter, the variable of participation in the maintenance of collective water conservancy project facilities is still selected from the perspective of output method, and the variable of participation in the decision-making of collective water conservancy project meeting is selected from the perspective of process method. However, there are some differences from the previous chapter, that is, the two indicators are taken as observed variables and combined into latent variables.

(3) Mediating variables

To explore the mechanisms that mediate how land fragmentation affects irrigation collective action, this book defined four mediating factors according to the questionnaire and the foregoing analysis.

In recent years, in China, the proportion of agricultural income in total

CHAPTER 6
MEDIATING EFFECT OF FARMLAND FRAGMENTATION ON RURAL IRRIGATION COLLECTIVE ACTION

rural household income has been declining; the Chinese economy has been changing over time from full-time agricultural work to part-time agricultural work, and then to non-agricultural work. From 2003 to 2016, the proportion of full-time agricultural households, part-time agricultural households, and non-agricultural household changed from 11%, 56%, and 33%, to 3%, 74%, and 23%, respectively (Zhang, 2019). This is the main reason we consider agricultural dependency as a key mediating factor: if the household's income comes mainly from farming, it is coded as 1, otherwise it is 0.

Rule-making is important aspect of irrigation governance. In China, the township is the basic administrative organization, while villages are autonomous self-governing units. Village organizations are authorized to make rules for irrigation governance, including rules for water distribution, management, and payment. Whether villages exercise these powers to make standardized rules for irrigation governance is an important mediating factor for how land fragmentation affects collective action. Following the literature, this book hypothesizes that self-governing villages are associated with high levels of collective action. This book codes the variable as 1 if there a unified irrigation rule in the village, and 0 otherwise.

In terms of economics, farmers face great difficulty because of their small scale and fragmented land holdings. This book speculates that farmers are motivated to engage in collective action as a mechanism for risk pooling. This book therefore chooses economic pressure as a potential mediating factor. This is coded as 1 if farmers face a lot of economic pressure to invest in maintaining irrigation facilities, and 0 otherwise.

With rapid urbanization in China in recent years, land circulation has gradually become common in rural areas. By the end of 2016, more than 35% of the arable land in rural China had been converted to other uses (Rural Development Institute Chinese Academy of Social Sciences, 2017).

The rate of land circulation will continue to increase alongside urbanization. Therefore, this book considers land circulation as a mediating factor that affects collective action. If the rural household is engaged in transfer of use rights, it is coded as 1, otherwise it was 0.

(4) Control variables

Based on the institutional analysis and development framework, control variables affecting irrigation collective action were selected from four perspectives in this book. For institutional context, two variables, stability of property rights and failure of village governance, were selected. For biophysical conditions, this book used five: distance to city, village water resources, farmland location, village topography, and farmland water scarcity. For attributes of community, this book used village size and village economic development. For household attributes, this book used family size, and age and education of the household head.

6.3.3 Model

Based on the basic SEM formula—the two equations specifying the relationship between the latent variables and the observed variables—this book add mediating factors to explore the mechanism by which farmland fragmentation influences irrigation collective action. The model (6-4) was constructed for the total effect of farmland fragmentation on irrigation collective action:

$$ICA_i = \alpha_0 + \alpha_1 \cdot LF_i + \sum \alpha_2 \cdot X_i + \varepsilon_1 \quad (6-4)$$

where *ICA* and *LF* are farmland fragmentation and irrigation collective action, respectively; and *X* represents the control variables that affect irrigation collective action, including stability of property rights, failure of village governance, distance to city, village water resource, farmland location, village topography, farmland water scarcity, village size, village economic development, family size, and household head age and education.

The effect of farmland fragmentation on irrigation collective action depends on whether the coefficient α_1 is significant. If it is significant, the effect of the mediating mechanism can be calculated by constructing the following models.

$$M_i = a_0 + a_1 \cdot LF_i + \varepsilon_2 \quad (6-5)$$

where M represents the mediating factors: dependency on farming, rule-making, economic pressure, and farmland ownership transfer. Equation (6-5) represents the influence of farmland fragmentation on the mediating factors. Then, the equation (6-6) represents the influence of farmland fragmentation, mediating factors, and external factors on irrigation collective action:

$$ICA_i = c_0 + c_1 \cdot LF + b_1 \cdot M_i + \sum c_2 \cdot X_i + \varepsilon_3 \quad (6-6)$$

Combining the equation (6-5) and equation (6-6) and gives equation (6-7), which indicates the direct influence of land fragmentation on irrigation collective action and the indirect influence of farmland fragmentation on irrigation collective action through mediating factors:

$$ICA_i = c_0 + a_0 \cdot b_1 + c_1 \cdot LF + a_1 \cdot b_1 \cdot LF_i + \sum c_2 \cdot X_i + b_1 \cdot \varepsilon_2 + \varepsilon_3$$
$$(6-7)$$

where c_1 is the direct influence coefficient and $a_1 b_1$ is the indirect influence coefficient.

6.3.4 Samples

The sample used in this part is the same as that used in chapter 5, including 3,895 peasant households in 284 villages in 17 provinces. Therefore, the descriptive statistical analysis of the data is consistent with chapter 5. But at the same time, the indicator of single farmland area is introduced in this chapter. Through analysis, it is found that the average size of a single plot of farmland for a farmer's family is only 0.09 ha, with a maximum value of 8.53 ha and a minimum value of 0.0067 ha. This further fully illustrates that the degree of farmland fragmentation owned by peasant households in China is more serious.

6.4 Total Effect of Farmland Fragmentation on Collective Action

As used in this book, both farmland fragmentation and irrigation collective action are latent variables, and each includes two observed variables, so the author wants to know whether each observed variable can effectively explain the latent variable. To do this, here use average variance extracted and composite reliability (see Table 6-2). They are greater than 0.5 and 0.7, respectively, which indicates that the observed variables do indeed explain the latent variables.

Table 6-2   Reliability statistics of the latent variables

Latent variables	Index variables	Average variance extracted	Composite reliability
Land fragmentation	Number of plots Average plot size	0.557	0.708
Irrigation collective action	Frequency of participation in collective maintenance Frequency of attending village meetings related to irrigation	0.619	0.743

In chapter 5, it has been verified that the total effect of farmland fragmentation on rural irrigation collective action is relatively significant, which can further explore the internal mechanism of farmland fragmentation affecting rural irrigation collective action. Because the total effect analysis results are significant, the mechanisms by which land fragmentation affects irrigation collective action can be further explored. On the basis of the SEM with mediating effect and the data from 3,895 rural households, we first check the degree of fit of the model. After calculation and model modification by Amos, we have an RMSEA of 0.049, just under the standard value of 0.05. This indicates that the model is well adapted to the sample data. In addition, GFI = 0.968, AGFI = 0.947, NFI = 0.861, IFI = 0.873, and CFI = 0.872; these are all greater than 0.8, indicating very good fit. Table 6-3 gives the estimates of the direct and indirect effects of land fragmentation on irrigation collective action.

Table 6-3 Result of direct and indirect effect of land fragmentation on irrigation collective action

Mechanism	Direct effect	Indirect effects		
		Influence (coefficient) of land fragmentation on mediating factor	Influence (coefficient) of mediating factor on irrigation collective action	Influence (coefficient) of land fragmentation on irrigation collective action based on mediating factor
Land fragmentation → Agricultural dependency → Irrigation collective action	-0.007	-0.158 ***	0.083 ***	-0.013 ***
Land fragmentation → Rule-making → Irrigation collective action		-0.283 ***	0.110 ***	-0.031 ***
Land fragmentation → Economic pressure → Irrigation collective action		0.044 ***	-0.147 ***	-0.006 ***
Land fragmentation → Land circulation → Irrigation collective action		0.081 ***	0.044 **	0.004 *
Sum of indirect effects				-0.047 ***

Notes: ***, **, and * indicate significance at the 0.5%, 2.5%, and 5% level, respectively.

FARMLAND FRAGMENTATION AND COLLECTIVE ACTION: A STUDY ON THE IRRIGATION SYSTEM IN CHINA

After the total effect of farmland fragmentation on irrigation collective action is divided into direct effect and indirect effects, the estimated direct effect is -0.007, which is very small, but also it is not statistically significant. Although farmland fragmentation could plausibly reduce the likelihood of cooperation in irrigation collective action, the author do not see such a direct influence in our results; farmland fragmentation seems to affect irrigation collective action mainly through mediating factors. That is to say, the conclusion that farmland fragmentation weakened the likelihood to cooperate in irrigation collective action mainly played a role on the basis of the mediating factors.

The estimated coefficients for agricultural dependency, rule-making, economic pressure, and farmland ownership transfer are all significant. Specifically, farmland fragmentation has a significant negative impact on both agricultural dependency and rule-making. That is, where farmland fragmentation is worse, farmers are less dependent on agricultural resources, so it is difficult to come to unified irrigation rules. Agricultural dependency and rule-making have significant positive effects on irrigation collective action. This means that peasants' dependency on agricultural resources and irrigation rule-making in rural areas are conducive to the improvement of irrigation collective action. Thus, farmland fragmentation could weaken the likelihood of cooperation in irrigation collective action through agricultural dependency and rule-making. These results are consistent with hypotheses H1 and H2.

In contrast, economic pressure presents a completely different process, although farmland fragmentation does have a significant negative impact on irrigation collective action through economic pressure. That is, the serious the farmland fragmentation, the more economic pressure there is on peasants in the construction and maintenance of agricultural irrigation facilities. But the greater the economic pressure, the less able peasants are to engage in

CHAPTER 6
MEDIATING EFFECT OF FARMLAND FRAGMENTATION ON RURAL IRRIGATION COLLECTIVE ACTION

irrigation collective action. This is consistent with hypothesis H3.

In the four mediating factors, farmland ownership transfer was the only one that presented different effects. When farmland fragmentation is relatively serious, it contributes to farmland circulation, which then promotes the improvement of irrigation collective action. Thus, farmland fragmentation increases irrigation collective action by increasing farmland ownership transfer. This is consistent with hypothesis H4.

Figure 6-1 sums up the role of the four mediating factors in the effect of farmland fragmentation on irrigation collective action.

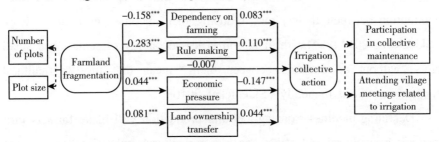

Figure 6-1 Mechanisms mediating the influence of farmland fragmentation on irrigation collective action

6.5 Discussion

The empirical results indicate that four mediating factors-dependency on farming for livelihood, formulation of rules for irrigation governance, economic pressure, and farmland ownership transfer-play important mediating roles in how farmland fragmentation influences irrigation collective action.

6.5.1 Effects of Dependency on Farming

Land fragmentation could have a negative effect on irrigation collective action through households' dependency on farming. This would happen if land fragmentation reduced farmers' profits from agricultural production and thus reduced their enthusiasm for participating in irrigation collective action. Three considerations might explain why land fragmentation would reduce dependency on farming.

The first reason is inefficient land utilization. If there are too many plots, boundary roads and irrigation facilities will occupy large areas which cannot produce any economic benefits because they cannot be planted (Di Falco, 2010). The second reason is inefficient time utilization. Managing fragmented land is more costly, because farmers spend a lot of time commuting between their plots, reducing their income per hour of work (Sklenicka, 2014). The third reason is less economy of scale. The small plots, without good roads between them, mean that most agricultural production is done manually (Sklenicka, 2016).

Declining incomes from agricultural production will make farmers turn to other agricultural industries, or non-agricultural industries, to increase their income, thus reducing their dependency on agricultural production, and thus eventually reducing the demand for agricultural irrigation facilities and the willingness to engage in irrigation collective action.

6.5.2　Effects of Irrigation Rule-making

Two reasons can be proposed for why land fragmentation has a negative effect on irrigation collective action, mediated by irrigation rule-making. The first reason is the difficulty of establishing unified irrigation rules. The more households there are, with their diverse irrigation demands, the less likely it will for everyone to reach an agreement on many items, such as water allocation, water price, and priority (Qiao, 2016).

The second reason is unfair sharing of collective action. Irrigation rules govern investment, construction, and maintenance of irrigation facilities, implying significant investments in capital, labor, and time (Guo, 2016). Land fragmentation, and the great number of households involved, all with different needs, make unfair division of labor more likely, and agreement between different households less likely (Miao, 2014). Thus, land fragmentation could reduce the likelihood of cooperation in irrigation collective action through the mediating factor of rule-making.

6.5.3　Effects of Economic Pressure

The survey results also imply that economic pressure mediates a negative effect of land fragmentation on irrigation collective action. Three reasons for this can be proposed.

First, land fragmentation increases the capital investment needed for agricultural production. Greater spatial dispersion of its land plots will require each household to construct more infrastructure, such as buildings on each plot and roads between the plots (Heider, 2018; Latruffe, 2014).

Second, greater land fragmentation means greater investments for agricultural machinery and equipment—requiring either multiple machines for multiple plots, or transporting the machines between plots, with the attendant costs in time and fuel (Abdollahzadeh, 2012).

Third, land fragmentation increases the organizational, management, and transaction costs of agricultural production. The more plots a farmer has, the more work is needed to optimize the allocation of resources (organization and management). This will make the farmer less likely to engage in irrigation collective action. Thus, land fragmentation reduces irrigation collective action through the mediating factor of economic pressure.

6.5.4　Effects of Land Circulation

Farmland fragmentation increases land circulation because it reduces agricultural production efficiency. This effect can be described with respect to three different types of efficiency.

The first is scale efficiency. Smaller plots prevent economies of scale. Farmers turn to land circulation to combine plots and recover economies of scale (Latruffe, 2014).

The second is production efficiency. More space between plots makes it harder to allocate resources among them, limiting efficiency. Land circulation ameliorates this problem as well.

The third is the utilization efficiency of machinery. Land fragmentation makes the utilization of machinery less efficient. For example, when crops are harvested, there are more losses in border and corner areas (Lu, 2018; Huy, 2017). The transfer of property rights will improve this kind of efficiency. Whether the property right is transferred out or transferred in, the purpose of land circulation is to improve efficiency and realize economies of scale in agricultural production. This implies a greater demand for irrigation facilities. Thus, land fragmentation leads to land circulation, which leads to cooperation in irrigation collective action.

6.6 Conclusion

Based on the SES framework and the data of 3,895 households in 284 villages in 17 provinces of China, this part studies the mechanism of the impact of farmland fragmentation on rural irrigation collective action by using structural equation model. The main innovation of this part is to make up for the influence of the key factor of farmland fragmentation in the current researches on the influencing factors of rural collective action, and to reveal the mediating mechanism through which this factor finally has an

CHAPTER 6
MEDIATING EFFECT OF FARMLAND FRAGMENTATION ON RURAL IRRIGATION COLLECTIVE ACTION

impact. Through the study, this book found that the farmland fragmentation reduces the farmers' ability to engage in the rural irrigation collective action, and this negative effect is not generated through the direct relationship between the two, but through many mediating factors such as agricultural dependence, system formation, economic pressure, land transfer, etc. Although different mediating factors play a role between the two presents the opposite relationship, but negative impact in the overall relationship, namely the farmland and finely by agricultural dependency, system formation, economic pressure, the factors of land circulation mediating role, inhibition of rural irrigation collective action in the whole.

CHAPTER 7

CONCLUSIONS, POLICY RECOMMENDATIONS AND PROSPECTS

7.1 Conclusions

Based on the practical background of the decline of collective action capacity of rural irrigation in China, this book explores the mechanism of the impact of farmland fragmentation on collective action capacity of rural irrigation from the perspective of the serious degree of farmland fragmentation with Chinese characteristics. First, based on the IAD framework, this book starts from the fundamental realities of a big country with a large population of smallholders, discusses the mechanism of small-scale peasants in the process of participating in China's rural irrigation collective action, and analyzes the influence of different characteristics of small-scale peasants on rural irrigation collective action. On this basis, it analyzes the results and conclusions of the current research from the perspectives of different farmland

CHAPTER 7
CONCLUSIONS, POLICY RECOMMENDATIONS AND PROSPECTS

attributes such as farmland size, farmland location and farmland property right, and then puts forward the main perspectives and entry points of this book.

Second, starting from the village level, this book constructs an econometric analysis model to analyze the impact of the shortage of per capita farmland resources in China on the rural irrigation collective action from the perspective of farmland resource endowment, and points out that although China's farmland resource endowment is insufficient, there is still a research conclusion that the stronger the degree of farmland resource endowment, the higher the collective action capacity of rural irrigation. Besides, this book discusses the total effect of farmland fragmentation on rural irrigation collective action from the aspect of small-scale peasants, and analyzes the regulating role of different institutional rules factors in the impact of farmland fragmentation on rural irrigation collective action. This book points out that the phenomenon of farmland fragmentation in China reduces the formation of rural irrigation collective action, but different institutional factors play an important role in regulating the relationship between the two.

Third, this book continues to explore the mediating mechanism of farmland fragmentation affecting rural irrigation collective action from the perspective of peasant-level. It points out that the phenomenon of farmland fragmentation does not directly affect the rural irrigation collective action in China, but acts through different mediating factors, and generally appears to inhibit the formation of the phenomenon of rural irrigation collective action. Specifically, the farmland fragmentation in China has reduced peasants' dependence on agricultural production, weakened the formation of rural irrigation system, and increased the pressure faced by farmers to engage in rural irrigation collective action. These phenomena have resulted in a serious decline in China's rural irrigation collective action capacity. However, the factor of farmland ownership transfer can play a positive role in the whole process, but its influence degree is far lower than the sum of other factors.

Overall, Figure 7-1 shows the main conclusions of this book.

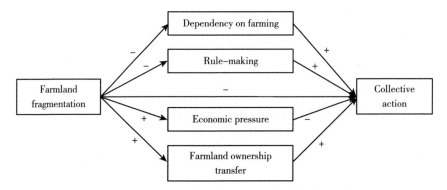

Figure 7-1 The logic diagram of farmland fragmentation affecting rural irrigation collective action

7.2 Policy Recommendations

In general, the features of farmland fragmentation go against the development of rural collective action in China. Especially, since the establishment and implementation of the Household Contract Responsibility System in the 1980s, the insufficient per capita resource endowment of small-scale peasants and the serious farmland fragmentation have gradually led to the continuous decline of rural collective action capacity in China. However, in the long run, the acceleration of farmland ownership transfer, agricultural production and management mode innovation and many other factors will contribute to the success of China's rural collective action, the future of China's rural collective action capacity will gradually enhance.

In a long period of time in the future, China's fundamental realities of a

CHAPTER 7
CONCLUSIONS, POLICY RECOMMENDATIONS AND PROSPECTS

big country with a large population of smallholders will not change, and small-scale peasants will serve as an important basic force to promote the development of China's agriculture and rural areas. According to the *National Population Development Plan* (2016 – 2030), China's total population will peak around 2030, about 1.45 billion. If the urbanization rate is calculated at 70%, China's rural population will reach 435 million in 2030. For China in the future, how to achieve effective rural governance in rural areas, and then achieve national harmony and stability, is a major issue now and in the future, but also a huge challenge to other countries. Therefore, in view of the current fundamental realities of China's a big country with a large population of smallholders and the current situation of the decline of rural collective action capacity, this book puts forward the following policy suggestions to solve the dilemma of China's rural collective action.

First, deepen reform of the rural land system. On the basis of implementing the policy of extending the second round of land contracts for another 30 years after their expiration, it is suggested to continue to deepen reform of the land system, fully safeguard the rights and functions of small-scale peasants in the process of land allocation and adjustment, and give full play to the rights of small-scale farmers in the development of rural communities while ensuring ownership, stabilizing contracting rights and activating management rights. At the same time, with the emergence of rural labor outflow and other phenomena, the rural land system also needs to constantly adjust, adapt and innovate based on the change of the actual situation, to ensure the fairness of each small-scale peasant in the rural community.

Second, deepen reform of the agricultural operation system. Continue to deepen reform of the agricultural operation system, vigorously promote new modes of agricultural production and operation, such as share-based land cooperation, trustee-holding for farming instead of farming, combined agricultural machinery, and socialized agricultural services. Gradually promote large-scale agricultural production and operation, constantly divide and eliminate small-scale peasants, and effectively increase the degree to which rural households engage in rural collective action. In addition, accelerate the process of farmers' professionalization to improve the self-development ability of small-scale peasants, so as to enhance the participation of farmers in the governance of rural public affairs.

Third, vigorously promote the development of rural industries. Industrial Prosperity is the top priority of the overall goal of the Rural Revitalization Strategy, and it is also the basis for the effective realization of the goal of Effective Governance. Adopt the mode of innovation of rural industrial development to promote the development of rural industry integration, promote the development of rural industry elements flow, constantly improve the level of rural industry development, and enhance the level of peasants' income and improving the capacity of small-scale scales participate in the supply of rural public goods, and ensure that the small-scale scales have more autonomy, thus improving the rural collective action capacity.

7.3　Prospects

7.3.1　Innovations

In this book, the effect of farmland fragmentation on peasants' participation in irrigation collective action has been analyzed by using field survey data. This book aims to reveal the effect of farmland fragmentation on rural irrigation collective action, and analyze the reasons behind it, so as to find the correct way for effective cooperation. In general, this book has the following innovations:

First of all, it enriches the existing theories related to collective action research, especially the IAD framework and SES framework. At present, more scholars have studied the rural irrigation collective action by using IAD framework and SES framework, and explained the influence of more than 30 external variables on the rural irrigation collective action. Although there are many literatures on the influence of farmland resources on irrigation collective action, the internal structural characteristics of farmland resources have not been specifically divided. In combination with the serious fragmentation of rural farmland in China, this book focuses on the variable with Chinese characteristics of farmland fragmentation to study the impact of farmland fragmentation on irrigation collective action, which enriches the IAD framework and SES framework and is a major innovation in the theoretical research of collective action.

Second, the mechanism of farmland fragmentation affecting rural irrigation collective action has been revealed. Compared with other countries in the world, the total area of farmland in China is larger, but due to the large population, the level of farmland per capita is at the bottom of the world. Although existing studies have shown that farmland fragmentation has an important effect on agricultural production, the relationship between farmland fragmentation and irrigation collective action is not clear. Therefore, the study on the key factor of farmland fragmentation, through the combination of qualitative analysis and quantitative analysis, reveals the logical relationship and the main mechanism of farmland fragmentation used in rural irrigation collective action, and establishes a logical diagram of the interaction between the two, which opens up a path for improving the collective action capacity of rural China.

Third, the potential mechanism is analyzed and verified by using big data, generating convincing results. This book is based on the relevant data of nearly 4,000 sample peasants from more than 10 provinces in China, covering a wide range of areas. Through the research with the help of big data, the research results can fully reflect the current impact mechanism of farmland fragmentation on rural irrigation collective action in China, which has a strong applicability.

7.3.2 Deficiency and Prospect

Due to the limitation of the research time and other aspects, there are still some deficiencies in the specific content of this book, which need to be further analyzed. First of all, the research indicators have certain limitations. Because the questionnaire design did not fully consider the various influencing factors of rural irrigation collective action, this book has selected some variables as control variables, but strictly speaking, these variables are not enough to explain the external influencing factors of rural irrigation collective action. Therefore, in the future research on this issue, it is still necessary to consider the various factors of rural irrigation collective action and improve the relevant research indicators. Second, the research method has some limitations. For example, when evaluating the total effect of farmland fragmentation on rural irrigation collective action, only the number of farmland patches is selected as the evaluation indicator of farmland fragmentation, which means that farmers have several disconnected plots of farmland with a small area. Therefore, only one indicator cannot fully reflect this feature of farmland fragmentation. However, due to the need to consider the regulatory effect of various institutional rules in this part, only existing research methods can be selected for evaluation and analysis. In the future analysis, the author may continue to explore the use of other methods, strive to reflect the characteristics of farmland fragmentation from multiple perspectives, and evaluate the regulatory effect of multiple policies. Finally, the mediation mechanism has not been fully considered. The external effects of farmland fragmentation have multiple angles and the factors that affect the rural irrigation collective action are also diverse. How can we identify the common medium between the two, and then judge the linkage effect produced by the medium? Although this book has tried to dig out four factors through the method of literature mathematics and physics, it is not enough to explain them well. Therefore, in the future research, it is still necessary to explore the mediating role between the two, and deeply analyze whether farmland fragmentation also affects the rural irrigation collective action through other media.

REFERENCE

[1] Abdollahzadeh, G., Kalantari, K., Sharifzadeh, A., Sehat, A. Farmland fragmentation and consolidation issues in Iran: an investigation from landholder's viewpoint. Journal of Agricultural Science and Technology, 2012, 14, 1441 −1452.

[2] Abubakari, Z., Molen, P. Van Der, Bennett, RM., Kuusaana, ED. Land consolidation, customary lands, and Ghana's Northern Savannah Ecological Zone: An evaluation of the possibilities and pitfalls. Land Use Policy, 2016, 54, 386 −398.

[3] Agrawal, A., Gupta, K. Decentralization and participation: the governance of common pool resources in Nepal's Terai. World Development, 2005, 27, 629 −649.

[4] Agrawal, A., Yadama, G. How do local institutions mediate market and population pressures on resources? Forest panchayats in Kumaon, India. Dev. Change, 1997, 28, 435 −465.

[5] Araral, E. The strategic games that donors and bureaucrats play: an institutional rational choice analysis. J. Publ. Adm. Res. Theory, 2008, 19, 853 −871.

[6] Araral, E. What explains collective action in the commons? Theory and evidence from the Philippines. World Development, 2009, 37, 687 − 697.

[7] Bai, Z., Chen, Y., Xie, B., et al. The relationship between landscape fragmentation and cultivated land use efficiency: Case of Kangle country, Gansu province. Journal of Arid Land Resources and Environment, 2014, 28 (4): 42 −47.

[8] Bardhan, P. Distributive conflicts, collective action, and institutional economics. Frontiers of Development Economics—The Future in Perspective, 2001, 7, 269 – 290.

[9] Binns, B. The consolidation of fragmented agricultural holdings. Food and Agricultural Organization, 1950.

[10] Cai, J., Ke, Y. Regional comparison of farmers' willingness of irrigation cooperation and its influencing factors: Based on the empirical analysis of 430 farmers in 15 provinces of China. China Rural Survey, 2015 (5): 62 – 72.

[11] Cai, Q., Zhu, Y. Analysis on farmers' willingness to participate in rural public goods supply. Journal of South China Agricultural University (Social Science Edition), 2014 (3): 45 – 51.

[12] Cai, Q., Zhu, Y. The impact of social capital and income gap on village collective action: a case study of farmers' participation in the maintenance of small irrigation and water conservancy facilities in three provinces. Journal of Public Management, 2016 (4): 89 – 100.

[13] Cai, Q., Zhu, Y. The influence of relationship network on farmers' participation in village collective action: a case study of farmers' participation in small-scale irrigation and water conservancy construction investment. Journal of Nanjing Agricultural University (Social Sciences Edition), 2017 (1): 108 – 118.

[14] Cai, R., Cai, S. Village size, income inequality and village collective action: a case study of 102 villages in Anhui province. Economic Review, 2014 (1): 48 – 57.

[15] Cai, R., Ma, W., Guo, X. Empirical analysis on farmers' willingness of cooperative supply of small irrigation and water conservancy facilities: a case study of farmland irrigation canal reconstruction in Yancheng City. Resources Science, 2014 (12): 2594 – 2603.

[16] Cai, R. Management and protection effect and Investment Willingness:

Analysis on the dilemma of cooperative supply of small irrigation and water conservancy facilities. Journal of Nanjing Agricultural University (Social Sciences Edition), 2015 (4): 78 –86.

[17] Chun, N. The challenge of collective action for irrigation management in India. Asian Econ. Pap., 2014, 13, 88 –111.

[18] Dai, S., Wu, B., Wu, D., et al. Study on priority of agricultural land consolidation in hilly areas based on fragmentation and location conditions: a case study of Songxi county, Fujian province. Journal of Guizhou Normal University (Natural Sciences), 2015, 33 (4): 14 – 20.

[19] Deininger, K., Savastano, S., Carletto, C. Land fragmentation, cropland abandonment, and land market operation in Albania. World Development, 2012, 40 (10): 2108 –2122.

[20] Demetriou, D., Stillwell, J., See, J. A new methodology for measuring land fragmentation. Computers, Environment and Urban Systems, 2013, 39 (5): 71 –80.

[21] Deng, Y., Xin, G., Wang, J., et al. Comprehensive evaluation of land fragmentation in hilly areas. Journal of Southwest University (Natural Science Edition), 2017 (3): 157 –163.

[22] Di, S., Penov, I., Aleksiev, A., Rensburg, TM. Van. Agrobiodiversity, farm profits and land fragmentation: Evidence from Bulgaria. Land Use Policy, 2010, 27 (3): 763 –771.

[23] Dijk, T. Scenarios of Central European land fragmentation. Land Use Policy, 2003, 20 (2): 149 –158.

[24] Ding, D., Zheng, F., Wu, L., et al. Economic and social heterogeneity and rural collective action level: Based on the data of 400 farmers in 40 villages of S County, Hubei Province. China Population, Resources and Environment, 2013 (9): 56 –61.

[25] Ding, J., Huang, G. Solving the dilemma of water conservancy and

irrigation through farmers' land cooperation: a case study of Wangping village in Shayang, Hubei Province. Rural Economy, 2012 (8): 122 – 125.

[26] Dirimanova, V. The impact of land fragmentation in ownership on farmland use in Bulgaria. Journal of Environmental Protection and Ecology, 2010, 11 (3): 1105 – 1110.

[27] Farley, KA., Ojeda-revah, L, Atkinson, EE., Eaton-gonzález BR. Changes in land use, land tenure, and landscape fragmentation in the Tijuana River Watershed following reform of the ejido sector. Land Use Policy, 2012, 29 (1): 187 – 197.

[28] Fitz, D. Evaluating the impact of market-assisted land reform in Brazil. World Development, 2018, 103, 255 – 267.

[29] Frija, A., Zaatra, A., Frija, I., AbdelHafidh, H. Mapping social networks for performance evaluation of irrigation water management in dry areas. Environ. Model. Assess, 2017, 22, 147 – 158.

[30] Fujiie, M., Hayami, Y., & Kikuchi, M. The conditions of collective action for local connoms management: The case of irrigation in the Philippines. Agricultural Economics, 2005, 33, 179 – 189.

[31] Gao, R., Wang, Y., Chen, C. Labor outflow and rural public affairs governance. China Population, Resources and Environment, 2016 (2): 84 – 92.

[32] Guo, G., Ding, C. Quantitative research on the impact of land fragmentation on grain production returns to scale: Based on the empirical data of Yancheng City and Xuzhou City in Jiangsu Province. Journal of Natural Resources, 2016 (2): 202 – 214.

[33] Guo, H., Li, Y., Gan, C., et al. Spatial characteristics of cultivated land fragmentation in the hinterland of the Three Gorges Reservoir Area: a case study of Fengjie County. Journal of Chongqing Normal University (Natural Science), 2016 (3): 51 – 58.

[34] Guo, Z. Farmland transfer, collective action and the supply of small irrigation and water conservancy facilities in villages: a case study of Tuanjie village in Hunan Province. Issues in Agricultural Economy, 2015 (8): 21 – 27.

[35] Hanemann, M. Property rights and sustainable irrigation—a developed world perspective. Agric. Water Management, 2014, 145, 5 – 22.

[36] Hartvigsen, M. Land reform and land fragmentation in Central and Eastern Europe. Land Use Policy, 2014, 36, 330 – 341.

[37] Hausner, V. H., Brown, G., Lægreid, E. Effects of land tenure and protected areas on ecosystem services and land use preferences in Norway. Land Use Policy, 2015, 49, 446 – 461.

[38] He, J., Liu, R., Hu, X. Differences in farmers' willingness to invest in small-scale irrigation and water conservancy and its influencing factors——Based on the comparison between major grain producing areas and non major grain producing areas. Journal of Hunan Agricultural University (Social Sciences), 2014 (4): 1 – 6.

[39] Heider, K., Rodriguez Lopez, J. M., Garcia Aviles, J. M., & Balbo, A. L. Land fragmentation index for drip-irrigated field systems in the Mediterranean: A case study from Ricote (Murcia, SE Spain). Agricultural Systems, 2018, 166, 48 – 56.

[40] Hoogesteger, J. Normative structures, collaboration and conflict in irrigation; a case study of the Píllaro North Canal Irrigation System, Ecuadorian Highlands. Int. J. Commons, 2015, 9, 398 – 415.

[41] Hoogesteger, J. Social capital in water user organizations of the Ecuadorian Highlands. Hum. Organ., 2013, 72, 347 – 357.

[42] Huang, S., Chen, Y., Zhang, R., et al. Spatial correlation analysis of cultivated land fragmentation and agricultural economic level based on landscape index. Agricultural Research in the Arid Areas, 2015 (3): 238 – 244.

[43] Huang, Z. , Wang, J. , Chen, Z. Effects of non-agricultural employment, land transfer and land fragmentation on rice farmers' technical efficiency. Chinese Rural Economy, 2014 (11): 4 – 16.

[44] Huy, Q. N. Analyzing the economies of crop diversification in rural Vietnam using an input distance function. Agricultural Systems, 2017, 153, 148 – 156.

[45] Jacoby, H. G. , Li, G. , Rozelle, S. Hazards of expropriation: tenure insecurity and investment in rural China. American Economic Review, 2002, 92, 1420 – 1447.

[46] Janus, J. , Markuszewska, I. Land consolidation—A great need to improve effectiveness. Land Use Policy, 2017, 65 (6): 143 – 153.

[47] Jennewein, J. S. , Jones, K. W. Examining 'willingness to participate' in community-based water resource management in a transboundary conservation area in Central America. Water Policy, 2016, 18, 1334 – 1352.

[48] Ji, Y. , Wang, X. , Lu, W. , et al. Characteristics of agricultural labor force, fragmentation of cultivated land and socialized service of agricultural machinery. Research of Agricultural Modernization, 2016 (5): 910 – 916.

[49] Jöreskog, K. G. , Goldberger, A. S. Factor analysis by generalized least squares. Psychometrika, 1972, 37 (3): 243 – 260.

[50] Kadirbeyoglu, Z. , Ozertan, G. Power in the governance of common-pool resources: a comparative analysis of irrigation management decentralization in Turkey. Environ. Policy Gov. , 2015, 25, 157 – 171.

[51] Kawasaki, K. The costs and benefits of land fragmentation of rice farms in Japan. The Australian Journal of Agricultural and Resources Economics, 2010, 54, 509 – 526.

[52] Ke, X. , Huang, X. , Hu, T. Farmers' willingness to participate in small-scale irrigation and water conservancy construction and its

influencing factors: an investigation based on economically developed counties. Journal of Hunan Agricultural University (Social Sciences), 2015 (3): 65 –69.

[53] King, R., Burton, S. Land fragmentation: notes on a fundamental rural spatial problem. Progress in Human Geography, 1982, 6 (4): 475 –496.

[54] Kjelland, ME., Kreuter, UP., Clendenin, GA. Factors related to spatial patterns of rural land fragmentation in Texas. Environmental Management, 2007, 40 (2): 231.

[55] Krusekopf, C. C. Diversity in land tenure arrangements under the household responsibility system in China. China Econ. Rev., 2002, 13, 297 –312.

[56] Lam, NS., Cheng, W., Zou, L., et al. Effects of landscape fragmentation on land loss. Remote sensing of environment, 2018, 209 (5): 253 –262.

[57] Latruffe, L., Piet, L. Does land fragmentation affect farm performance? A case study from Brittany, France. Agric Syst. 2014, 129, 68 –80.

[58] Li, G., Zhong, F. Farmland fragmentation, labor force utilization and farmers' income: An empirical study based on underdeveloped areas in Jiangsu Province. Chinese Rural Economy, 2006 (4): 42 –48.

[59] Li, W., Zhang, L. Villagers' autonomy: collective action, institutional change and cultivation of public spirit: a case study of ganchangpo village group autonomy in Xishui County, Guizhou Province. Management World, 2008 (10): 64 –74.

[60] Li, X., Ou, M., Xiao, C., et al. Study on the impact of fragmentation on cultivated land productivity based on landscape index. Resources and Environment in the Yangtze Basin, 2012 (6): 707 – 713.

[61] Li, Y., Chen, C. Fragmentation, scale effect and input output

efficiency of rice farmers. Journal of South China Agricultural University (Social Science Edition), 2011 (3): 72 –78.

[62] Lian, X., Mao, Y., Wang, H. Property right, transaction cost and agricultural production of fragmented land: an empirical survey from Wucun village in the central plain of Inner Mongolia. China Population, Resources and Environment, 2014 (4): 86 –92.

[63] Lian, X., Mao, Y. Land fragmentation will inevitably lead to the reduction of land production efficiency: Critical analysis on land fragmentation and land productivity. Journal of Huazhong Agricultural University (Social Sciences Edition), 2013 (6): 109 –115.

[64] Liu, Q., Qu, W., Li, Z. Investigation and Analysis on the impact of farmland fragmentation on crop production and farmers' income in Arid Oasis Area: a case study of Minle County, Gansu Province. Agricultural Research in the Arid Areas, 2011 (3): 191 –198.

[65] Liu, T., Qu, F., Jin, J., et al. The impact of land fragmentation and land transfer on Farmers' land use efficiency. Resources Science, 2008 (10): 1511 –1516.

[66] Looga, J., Jürgenson, E., Sikk, K., Matveev, E., Maasikamäe, S. Land fragmentation and other determinants of agricultural farm productivity: the case of Estonia. Land Use Policy, 2018, 79, 285 – 292.

[67] Lu, H., Hu, H., Geng, X. Land fragmentation, plot size and agricultural production efficiency: Empirical analysis based on survey data of Jiangsu Province. Journal of Huazhong University of Science and Technology (Social Science Edition), 2016 (4): 81 –90.

[68] Lu, H., Hu, H. Analysis of the impact of land fragmentation and planting diversification on agricultural production profit and efficiency: Based on the micro survey of Jiangsu farmers. Journal of Agrotechnical Economics, 2015 (7): 4 –15.

[69] Lu, H., Hu, H. Does land fragmentation increase the cost of agricultural production: Micro survey from Jiangsu Province. Economic Review, 2015 (5): 129-140.

[70] Lu, H., Hu, H. Non-agricultural labor supply: does land fragmentation work: Based on the perspective of Lewis turning point. Economic Review, 2017 (1): 148-160.

[71] Lu, H., Xie, H., He, Y., et al. Assessing the impacts of land fragmentation and plot size on yields and costs: A translog production model and cost function approach. Agricultural System, 2018, 161 (1): 81-88.

[72] Lyu, X., Huang, X., Zhong, T., et al. Research progress of farmland fragmentation in China. Journal of Natural Resources, 2011, 4 (3): 530-540.

[73] Lyu, Z., Niu, L., Hao, J., et al. Study on multi index comprehensive evaluation of cultivated land fragmentation degree based on analytic hierarchy process. Chinese Agricultural Science Bulletin, 2014, 30 (26): 200-206.

[74] Manjunatha, A. V., Reza, A., Speelman, S., Nuppenau, EA. Impact of land fragmentation, farm size, land ownership and crop diversity on profit and efficiency of irrigated farms in India. Land Use Policy, 2013, 31, 397-405.

[75] Meinzen-Dick, R., DiGregorio, M., McCarthy, N. Methods for studying collective action in rural development. Agricultural Systems, 2004, 82, 197-214.

[76] Miao, S. Research on farmers' cooperative participation behavior in small water conservancy facilities from the perspective of multi-dimensional heterogeneity of social capital. China Population, Resources and Environment, 2014 (12): 46-54.

[77] Murtinho, F. What facilitates adaptation? An analysis of community-

based adaptation to environmental change in the Andes. Int. J. Commons, 2016, 10, 119 – 141.

[78] Mushtaq, S., Dawe, D., Lin, H., & Moya, P. An assessment of collective action for pond management in Zhanghe Irrigation System (ZIS), China. Agricultural Systems, 2007, 92, 140 – 156.

[79] Nagrah, A., Chaudhry, A. M., Giordano, M. Collective action in decentralized irrigation systems: evidence from Pakistan. World Development, 2016, 84, 282 – 298.

[80] Niroula, GS. Impacts and causes of land fragmentation and lessons learned from land consolidation in South Asia. Land Use Policy, 2005, 22, 358 – 372.

[81] Olson, M. The logic of collective action. Cambridge: Harvard University Press, 1965.

[82] Ostrom, E., Gardner, R., Walker, J. Rules, games, and common pool resources. Michigan: University of Michigan Press, 1994.

[83] Ostrom, E. Understanding institutional diversity. Princeton: Princeton University Press, 2005.

[84] Panagopoulos, Y., Makropoulos, C., Gkiokas, A., Kossida, M., Evangelou, L., Lourmas, G., Michas, S., Tsadilas, C., Papageorgiou, S., Perleros, V., Drakopoulou, S., Mimikou, M. Assessing the cost-effectiveness of irrigation water management practices in water stressed agricultural catchments: the case of Pinios. Agric. Water Manage, 2014, 139, 31 – 42.

[85] Patt, A. Beyond the tragedy of the commons: reframing effective climate change governance. Energy Res. Soc. Sci., 2017, 34, 1 – 3.

[86] Poteete, A. R., Janssen, M. A., Ostrom, E. Working together: Collective action, the commons, and multiple methods in practice. Princeton University Press, 2010.

[87] Poteete, A. R., Ostrom, E. Fifteen years of empirical research on

collective action in natural resource management: Struggling to build large-N databases based on qualitative research. World Development, 2008, 36, 176 – 195.

[88] Qiao, D., Lu, Q., Xu, T. Analysis on influencing factors of cooperative supply willingness of rural small water conservancy facilities based on multi group structural equation model. Rural Economy, 2016 (3): 99 – 104.

[89] Qin, L., Zhang, N., Jiang, Z. Land fragmentation, labor transfer and Chinese farmers' grain production: Based on the survey of Anhui Province. Journal of Agrotechnical Economics, 2011 (11): 16 – 23.

[90] Qiu, F., Laliberté, L., Swallow, B., Jeffrey, S. Impacts of fragmentation and neighbor influences on farmland conversion: A case study of the Edmonton-Calgary Corridor, Canada. Land Use Policy, 2015, 48, 482 – 494.

[91] Rahman, S., Rahman, M. Impact of land fragmentation and resource ownership on productivity and efficiency: The case of rice producers in Bangladesh. Land Use Policy, 2008, 26, 95 – 103.

[92] Ren, G. Problems and governance of farmers' participation in the management and protection of farmland water conservancy facilities: Based on the field survey of 420 farmers in southern Jiangsu. Journal of Hunan Agricultural University (Social Sciences), 2016 (3): 59 – 64.

[93] Ricks, J. I. Building participatory organizations for common pool resource management: water user group promotion in Indonesia. World Development, 2016, 77, 34 – 47.

[94] Sabates-Wheeler, R. Farm Strategy, Self-Selection and Productivity: Can Small Farming Groups Offer Production Benefits to Farmers in Post-Socialist Romania. World Development, 2002, 30 (10): 1737 – 1753.

[95] Schultz, T. W. Transforming Traditional Agriculture. Chicago:

University of Chicago Press, 1964.

[96] Sharaunga, S., Mudhara, M. Determinants of farmers' participation in collective maintenance of irrigation infrastructure in KwaZulu-Natal. Phys. Chem. Earth, 2018, 105, 265 –273.

[97] Shen, C., Feng, D., Wang, X., et al. Improvement on calculation of farmland fragmentation measurement index. Resources Science, 2012, 34 (12): 2242 –2248.

[98] Sheng, Y., Yao, Y., Qin, F., et al. Study on the classification of cultivated land fertility based on GIS. Journal of Arid Land Resources and Environment, 2014 (6): 27 –32.

[99] Sikor, T., Müller, D., Stahl, J. Land Fragmentation and Cropland Abandonment in Albania: Implications for the Roles of State and Community in Post-Socialist Land Consolidation. World Development, 2009, 37 (8): 1411 –1423.

[100] Sklenicka, P., Janovska, V., Salek, M., Vlasak, J., Molnarova, K. The Farmland Rental Paradox: Extreme land ownership fragmentation as a new form of land degradation. Land Use Policy, 2014;38, 587 – 593.

[101] Sklenicka, P., Zouhar, J., Trpáková, I., et al. Trends in land ownership fragmentation during the last 230 years in Czechia, and a projection of future developments. Land Use Policy, 2017, 67 (2): 640 –651.

[102] Sklenicka, P. Classification of farmland ownership fragmentation as a cause of land degradation: A review on typology, consequences, and remedies. Land Use Policy, 2016, 57, 694 –701.

[103] Su, X. Economic history of new China. CPC Central Committee Party School Press, 2007.

[104] Sun, Y., Liu, Y. Evaluation of sustainable land use based on fragmentation: a case study of Fenyi County, Jiangxi Province.

Journal of Natural Resources, 2010, 25 (5): 802 −810.

[105] Sun, Y., Zhao, X. Mesoscale study on land fragmentation in Fenyi County. China Land Science, 2010, 24 (4): 25 −31.

[106] Tai, X., Xiao, W., Zhang, J., et al. Study on cultivated land fragmentation in Chaohu basin based on landscape index. Research of Soil and Water Conservation, 2016, 23 (4): 94 −98.

[107] Tan, S., Heerink, N., Qu, F. Land fragmentation and its driving forces in China. Land Use Policy, 2006, 23, 272 −285.

[108] Thiesenhusen, W. C., Melmed-Sanjak, J. Brazil's agrarian structure: Changes from 1970 through 1980. World Development, 1990, 18 (3): 393 −415.

[109] Tian, M., He, X. Farmland fragmentation and its governance in China. Journal of Jiangxi University of Finance and Economics, 2015 (2): 88 −96.

[110] Totin, E., Leeuwis, C., van Mierlo, B., Mongbo, R., Stroosnijder, L., Kossou, D. Drivers of cooperative choice: canal maintenance in smallholder irrigated rice production in Benin. Int. J. Agric. Sustain., 2014, 12, 334 −354.

[111] Vidal-Macua, J. J., Ninyerola, M., Zabala, A., Domingo-Marimon, C., Gonzalez-Guerrero, O., Pons, X. Environmental and socioeconomic factors of abandonment of rainfed and irrigated crops in northeast Spain. Appl. Geogr., 2018, 90, 155 −174.

[112] Wang, D., Chen, Y., Jia, S., et al. Study on cultivated land fragmentation based on landscape pattern index: a case study of northern tableland in Linxia, Gansu Province. Chinese Agricultural Science Bulletin, 2014, 30 (32): 184 −188.

[113] Wang, H. Group characteristics and irrigation self-organized governance: a comparative study of two villages. Journal of Public Administration, 2013 (6): 82 −106.

[114] Wang, S., Feng, G. Research on China's agricultural land fragmentation and its integrated governance. Social Sciences in Yunnan, 2016 (1): 17 –22.

[115] Wang, X., Bi, R., Zhao, J. Research on influencing factors of farmland management fragmentation based on GWR: a case study of Dongjia village, Taigu County, Shanxi Province. Journal of Shanxi Agricultural University (Social Science Edition), 2016 (2): 133 – 138.

[116] Wang, X., Lu, Q. Empirical study on the willingness of cooperative supply of small water conservancy facilities in rural communities. China Population, Resources and Environment, 2012 (6): 115 –119.

[117] Wang, X., Lu, Q. Farmers' willingness to share the cost of cooperative supply of small water conservancy facilities and its influencing factors: Based on the survey data of Shaanxi Province. Journal of Huazhong Agricultural University (Social Sciences Edition), 2014 (5): 48 –52.

[118] Wang, X., Zhong, F. Farmland fragmentation and agricultural land transfer market. China Rural Survey, 2008 (4): 29 –34.

[119] Wang, Y., Wang, X. Canal irrigation management performance and its influencing factors in China. China Public Administration Review, 2014, 16 (2): 47 –68.

[120] Wang, Y. Improving rural collective action capacity and accelerating agricultural science and technology progress. Bulletin of Chinese Academy of Sciences, 2017 (10): 1096 –1102.

[121] Wang, Y. Reassessment of the framework of institutional analysis and development. China Public Administration Review, 2017 (1): 3 – 21.

[122] Wang, Y. Reform of China water users association: A perspective of policy implementation. Management World, 2013 (6): 61 –71.

[123] Wen, G., Yang, G., Li, Y., et al. Analysis on the effect of farmland consolidation on farmland fragmentation and its causes: a case study of Jiangxia, Xian'an and Tongshan districts (counties) in Hubei Province. China Land Science, 2016 (9): 82–89.

[124] Wen, G., Yang, G., Wang, W., et al. Evaluation of cultivated land fragmentation degree from the perspective of farmers: a case study of "Jiangxia District Xian'an District Tongshan County" in Hubei Province. Progress in Geography, 2016, 35 (9): 1129–1143.

[125] Wu, M., Li, R., Yang, Y. Impact of land production convergence on Farmers' purchase of agricultural machinery under the background of land fragmentation. Journal of Northwest A&F University (Social Science Edition), 2017 (2): 113–122.

[126] Wu, Q., Lin, H. Repeated game, community capacity and cooperation in irrigation and water conservancy. China Rural Survey, 2017 (6): 86–99.

[127] Wu, Y., Nie, Y., Hu, Z., et al. Family life cycle, land fragmentation and farmers' agricultural productive input: data from Laohekou City, Hubei Province. Journal of Yunnan University of Finance and Economics, 2008 (1): 70–75.

[128] Xia, Z. Logic and steps of innovation of farmland transfer system. Journal of South China Agricultural University (Social Science Edition), 2014 (3): 1–8.

[129] Xie, H., Lu, H. Land use policy impact of land fragmentation and non-agricultural labor supply on circulation of agricultural land management rights. Land Use Policy, 2017, 68 (4): 355–364.

[130] Xu, L., Luo, D., Liu, A. Impact of social capital on Farmers' willingness to participate in irrigation management reform. Resources Science, 2015 (6): 1287–1294.

[131] Xu, Q., Tian, S., Xu, Z., et al. Farmland system, land fragmentation

and farmers' income inequality. Economic Research Journal, 2008 (2): 83 –92.

[132] Xu, Q., Yin, R., Zhang, H. Economies of scale, returns to scale and agricultural moderate scale management: An empirical study based on China's grain production. Economic Research Journal, 2011 (3): 59 –71.

[133] Yang, L., Zhu, Y., Ren, Y. The influence of social trust and organizational support on the performance of farmers' participation in small-scale rural water conservancy. Resources Science, 2018, 40 (6): 1230 –1245.

[134] Yang, Z., Yang, G. Impact of farmland fragmentation on farmers' farmland transfer decision. China Land Science, 2017 (4): 33 –42.

[135] Ye, C., Xu, Q., Xu, Z. Causes and effects of farmland fragmentation: an economic explanation from a historical perspective. Issues in Agricultural Economy, 2008 (9).

[136] Yuan, J., Du, W. Influence factors for peasant households' behaviors to participate in management of small-scale irrigation facilities based on investigation of farmers in Fangcheng County of Henan Province. Journal of Economics of Water Resources, 2016 (3): 69 –74.

[137] Yucer, A., Kan, M., Demirtas, M., et al. The importance of creating new inheritance policies and laws that reduce agricultural land fragmentation and its negative impacts in Turkey. Land Use Policy, 2016, 56 (11): 1 –7.

[138] Yuko, N., Keijiro, O. Determinants of household contributions to collective irrigation management: a case of the Doho rice scheme in Uganda. Contributed Paper presented at the Joint 3rd AAAE and 48th AEASA Conference, Cape Town, South Africa. 2010, September, 19 –23.

[139] Zhang, C. , Peng, C. , Kong, X. Evolution logic, historical evolution and future prospect of farmer differentiation. Reform, 2019 (2).

[140] Zhang, H. , Yang, G. Farmland fragmentation and its impact on technical efficiency of grain production——Based on the production function of transcendental logarithm stochastic frontier and farmers' micro data. Resources Science, 2012, 34 (5): 903 –910.

[141] Zhang, H. Big country and small farmer: a historical choice for modernization. Seeker, 2019 (1).

[142] Zhang, Q. , Chen, W. , Luo, B. Farmland transfer, fragmentation improvement and transformation of agricultural management behavior: An empirical study based on the questionnaire of farmers in nine provinces. Rural Economy, 2017 (6): 1 –10.

[143] Zhang, Y. , Zhuo, J. Double demonstration of positive and negative effects of land fragmentation: An empirical study based on fixed observation data of farmers in Hebei Province. Journal of Jiangxi Agricultural University (Social Science Edition), 2008 (4): 25 –29.

[144] Zhong, F. , Wang, X. Can the current farmland transfer market reduce the degree of land fragmentation: Preliminary evidence from Xinghua County, Jiangsu Province and Binxian County, Heilongjiang Province. Issues in Agricultural Economy, 2010, 31 (1): 23 –32.

[145] Zu, J. , Zhang, B. , Kong, X. Characteristics and utilization efficiency of cultivated land fragmentation in Southwest Hilly Region: A case study of Caohai village in Guizhou Province. Journal of China Agricultural University, 2016, 21 (1): 104 –113.

图书在版编目（CIP）数据

耕地细碎化与农村集体行动：以中国农田灌溉系统为例＝Farmland Fragmentation and Collective Action：A Study on the Irrigation System in China：英文／臧良震著．—北京：经济科学出版社，2021.9
ISBN 978－7－5218－2844－3

Ⅰ.①耕… Ⅱ.①臧… Ⅲ.①农田灌溉－研究－中国－英文 Ⅳ.①S275

中国版本图书馆 CIP 数据核字（2021）第 182558 号

责任编辑：初少磊 李 宝
责任校对：徐 昕
责任印制：范 艳

耕地细碎化与农村集体行动：
以中国农田灌溉系统为例

Farmland Fragmentation and Collective Action：
A Study on the Irrigation System in China

臧良震 著

经济科学出版社出版、发行 新华书店经销
社址：北京市海淀区阜成路甲 28 号 邮编：100142
总编部电话：010－88191217 发行部电话：010－88191540
网址：www.esp.com.cn
电子邮箱：esp@esp.com.cn
天猫网店：经济科学出版社旗舰店
网址：http://jjkxcbs.tmall.com
北京季蜂印刷有限公司印装
710×1000 16 开 11 印张 190000 字
2021 年 11 月第 1 版 2021 年 11 月第 1 次印刷
ISBN 978－7－5218－2844－3 定价：45.00 元
（图书出现印装问题，本社负责调换。电话：010－88191510）
（版权所有 翻印必究 举报电话：010－88191586
电子邮箱：dbts@esp.com.cn）